EVOLUTION, RELIGION, AND COGNITIVE SCIENCE

Evolution, Religion, and Cognitive Science

Critical and Constructive Essays

EDITED BY
FRASER WATTS AND LÉON TURNER

OXFORD
UNIVERSITY PRESS

OXFORD
UNIVERSITY PRESS

Great Clarendon Street, Oxford, OX2 6DP,
United Kingdom

Oxford University Press is a department of the University of Oxford.
It furthers the University's objective of excellence in research, scholarship,
and education by publishing worldwide. Oxford is a registered trade mark of
Oxford University Press in the UK and in certain other countries

First Edition published in 2014

Impression: 1

Published in the United States of America by Oxford University Press
198 Madison Avenue, New York, NY 10016, United States of America

British Library Cataloguing in Publication Data
Data available

Library of Congress Control Number: 2013955759

ISBN 978–0–19–968808–1

Printed and bound by
CPI Group (UK) Ltd, Croydon, CR0 4YY

Preface

The Cognitive Science of Religion (CSR) has attracted a huge amount of attention from all quarters over many years, and has achieved something approaching notoriety in contemporary religious studies. It is already a very large and complex field of study, and continues to expand at a prodigious rate. However, although the development of CSR undoubtedly ranks amongst the most significant events in religious studies for many decades, we must be wary that the sheer missionary enthusiasm shown for it by its strongest adherents does not transform it into a quasi-religion in itself, and deflect careful critical analysis.

The main focus of this book is CSR in an evolutionary context—what has sometimes been called the evolutionary cognitive science of religion (ECSR). We mean neither to offer uncritical enthusiasm for ECSR, nor strongly oppose it, but rather to present careful, judicious, and detailed appraisals of it from many different perspectives. A recurrent theme of the various chapters is that, although they may have made many constructive contributions to the study of many aspects of religion, CSR and ECSR have followed a narrower path than they need have. Many of us want to take a 'broaden and build' approach to these disciplines, recognizing their contributions, but charting how they might develop in less constrained ways in the future. As the title of the book suggests, then, the essays presented here are frequently both critical and constructive. Collectively, they represent a broad range of academic disciplines, reflecting both the multi-disciplinary history of CSR, and the extent of its actual and potential impact upon religious studies more generally. We hope this book will serve as a guide to issues and debates within ECSR for students, researchers, and other non-specialists interested in the field.

Although some of the attention directed towards ECSR in recent years has resulted from its application to wider debates about religion, we wish to be quite clear that we have tried to approach the subject from a religiously neutral academic standpoint. We are not arguing for or against religion. One of the editors is religious, and the other is not; at no point in the editorial process has this been a relevant consideration.

The idea for the book arises from two conferences held in Cambridge in 2009, a double Darwin anniversary, being both the 200th anniversary of Darwin's birth and the 150th anniversary of the publication of *The Origin of Species*. The first of these was the wide-ranging Darwin Festival organized by the University of Cambridge, and the second was a conference of the International Society for Science and Religion. Both conferences included lengthy sessions on the evolution of religion. Both were supported by the John Templeton Foundation, as was the research of Fraser Watts and Léon Turner on the evolution of religion in the Psychology and Religion Research Group at the University of Cambridge. We are indeed most grateful to the Foundation for the support that has helped to make this book possible.

The book is, however, very far from being a conference proceedings, as less than half the chapters were presented as papers at those 2009 conferences, and those that were have been substantially revised and developed. During the long editing process, our vision for the book changed markedly, leading us to seek potential contributors from a range of fields that were not represented at the 2009 conferences. We are very grateful to Tom Perridge at OUP and several anonymous reviewers, whose advice strongly influenced the final preparation of the manuscript. Finally, we are enormously grateful to our contributors for their great patience with a rather protracted editorial process. It is our hope that the resulting sum is greater than the parts, and that the book as a whole may mark a step towards the maturity of ECSR as a field of open-minded scientific enquiry.

Fraser Watts and Léon Turner

Contents

List of Contributors

William Sims Bainbridge, Department of Sociology, George Mason University, and National Science Foundation, USA.

Nathaniel F. Barrett, Institute for Culture and Society, University of Navarra.

Omar Sultan Haque, Department of Psychology, Harvard University.

J. Wentzel van Huyssteen, Department of Theology, Princeton Theological Seminary.

Malcolm Jeeves, Emeritus Professor of Psychology, University of St Andrews.

Timothy Jenkins, Faculty of Divinity, University of Cambridge.

Lesley Newson, School of Psychology, University of Exeter.

Benjamin Grant Purzycki, Centre for Human Evolution, Cognition, and Culture, University of British Columbia.

Ilkka Pyysiäinen, Faculty of Theology, University of Helsinki.

Peter Richerson, Department of Environmental Science and Policy, University of California Davis.

Michael Ruse, Department of Philosophy, Florida State University.

Richard Sosis, Department of Anthropology, University of Connecticut.

Léon Turner, Faculty of Divinity, University of Cambridge.

Aku Visala, Department of Anthropology, University of Notre Dame.

Fraser Watts, Emeritus reader in Theology and Science.

1

Introduction: Pluralism and Complexity in the Evolutionary Cognitive Science of Religion

LÉON TURNER

In July 2009, two conferences took place in Cambridge (UK) at which the impact of evolutionary theory upon theology and religious studies were important themes. The first was a major international event marking the 150th anniversary of the publication of *The Origin of Species* and the 200th anniversary of Darwin's birth, and the second was organized by the International Society for Science and Religion. The variety of papers at both conferences demonstrated both the enduring significance of Darwinism for religious studies and the remarkable multiplicity of ways in which evolutionary theory might inform our understanding of religion. But it was also apparent, and this came as no surprise, that the evolutionary study of religion is increasingly dominated by those accounts that draw specifically upon the theories and data of the so-called cognitive science of religion (CSR). Those are the accounts that are the main focus of the essays comprising this book. Our primary aim is to explore various important ideas and assumptions of what Edward Slingerland and Joseph Bulbulia (2011) have recently called simply the 'evolutionary cognitive science of religion' (ECSR) from the perspectives of a wide range of relevant academic subjects. The contributors, several of whom delivered papers at one or other of the two Cambridge conferences in 2009, are each specialists in a range of different fields relevant to the study of religion, including theology, history and philosophy of science, evolutionary biology, cognitive neuroscience, psychology, social anthropology, sociology, artificial intelligence, and, of course, the cognitive science of religion.

In this brief introduction, we have two main aims. First, we mean to provide a glimpse of the great complexity and theoretical diversity of ECSR, and the range of criticisms it has attracted. Secondly, we will briefly outline the shape of the book and the nature of the contribution we hope it will make to the study of ECSR.

COGNITIVE SCIENCE AND RELIGION

At around a quarter of a century old, more or less, the CSR is still relatively young, but its influence upon the study of religion has already been very significant. CSR draws freely from work in a range of fields including, of course, cognitive psychology and religious studies, but also from the philosophy of mind, and neuroscience, as well as social, cultural and cognitive anthropology (for historical reviews of the early years of this field, see Atran 1993; Bulbulia 2004; Pyssiäinen (this volume); Slone 2004). With the support of these disciplines, CSR has covered a large number of topics including religious rituals (Lawson and McCauley 1990), belief in gods and spirits (Barrett 2004), belief in the afterlife (Bering 2011), magic (Sørensen 2006), notions of divine punishment and reward (Baumard and Chevallier 2012), modes of religiosity (Whitehouse 2004), and morality (Boyer 2002; Pyysiäinen and Hauser 2010), amongst other things. Despite the breadth of the field, its various strands are united by the belief that many aspects of religion are constrained to a certain extent by the normal functioning of the cognitive system, and the belief that the study of those constraints bolsters our understanding of religion in all its forms. That is to say, at least some (and perhaps most) religious beliefs and behaviours are assumed to be culturally underdetermined.

Beyond this basic claim about the constraining influence of everyday cognitive functioning upon religion, there is little truly universal consensus in the field, though some ideas are more firmly entrenched than others. Perhaps the best established, and certainly the most widely disseminated, are those that constitute what is increasingly commonly referred to as the 'standard model' in CSR. The key principles of this model are all clearly exhibited in one of the best developed and empirically supported of all areas of CSR research—the study of beliefs in supernatural agents. Although different theorists differ over the finer details, it is typically claimed that humans have strong unconscious predilections towards explaining quite a wide variety of phenomena in terms of the actions and intentions of minded supernatural agents, which are generated by what Justin Barrett has referred to as the 'maturationally natural cognitive system' (2010, p. 171), and Robert McCauley calls 'maturationally natural perceptual systems' (2011, p. 37). Supernatural agent concepts are assumed to arise from the normal operation of the cognitive system in response to certain challenges faced universally by human beings, and/or in response to certain kinds of ubiquitous environmental stimuli. In making this claim, CSR leans heavily upon a modular account of cognition. Essentially, this means the mind is not conceived as a general problem-solving device, but as constituted by a variety of subsystems, engaged in 'content-specific operations' (Pyysiäinen 2002, p. 1) that are proper to the particular cognitive domains with which they are concerned. CSR typically views these subsystems as more or less natural cognitive 'devices', specialized for the performance of particular tasks such as categorizing various objects according to specific 'native ontological categories' (Murray and Goldberg 2009, p. 184), or elaborating objects' properties once they have been appropriately categorized (Barrett 2004, p. 4). Ever vigilant, once triggered (and some of them are triggered very easily indeed), they cause all sorts of assumptions to be made about particular objects or events, which are appropriate for the objects/events of the category to which they are assigned.

Importantly, it is assumed there is no special cognitive faculty that deals exclusively with supernatural phenomena in general, or supernatural agent concepts, specifically. Supernatural agent concepts are taken to be products (and in at least one important sense, by-products) of normal cognitive functioning. But they differ from run-of-the-mill agent concepts in at least one very important respect—they violate some intuitive assumptions about agents in general. Supernatural agent concepts might then be seen, in Boyer's words, to have some specific features that violate some default expectations for the domain' and others that meet expectations 'held by default as true of the entire domain' (2003, p. 119). Crucially, however, such concepts are not so outrageous as to detract from their overall credibility—they are only 'minimally counterintuitive'. Actually, their relatively minor conceptual idiosyncrasies only serve to make them more interesting and more memorable. Once formed in an individual mind, they can spread easily and efficiently because they are so rich in 'inferential potential'—they are easily incorporated into complex, emotionally charged narratives, and readily supported and reinforced by natural cognitive resources.

The standard model has its detractors and revisionists, of course, but it is still aptly named. Notions of the largely innate, universal, functionally specialized cognitive tool-kit, the rich inferential potential and minimal counterintuitiveness shared by different sorts of religious ideas, the focus on supernatural agency, as well as an emphasis upon common features of how religious phenomena are represented and processed internally are all recurrent themes of CSR.

It is also clear that, in its concern with both cognition and culture, CSR is characterized by its focus upon universals rather than particulars, naturalness rather than social constructedness, and intuitive, unconscious influences rather than reasoned and conscious ones. McCauley and Whitehouse (2005, p. 2) see this as redressing an imbalance in other branches of religious studies. That is not to say CSR is completely blind to the particularity, social constructedness or conscious reasoned development of many aspects of religion.

But the primary concern of CSR is always fundamentally with those cognitive structures, processes and capacities that enable, or stand behind all particular expressions of religiosity. For CSR, this is the proper level at which to explain the ubiquity, persistence and, to a certain extent, the convergence of religious beliefs and behaviours all over the world.

EVOLUTIONARY COGNITIVE SCIENCE OF RELIGION: A MULTI-FACETED RESEARCH PROGRAMME

The cognitive turn in religious studies has proved particularly attractive to those seeking to explain the origins and subsequent development of religion *per se* in evolutionary terms. Awareness of the potential for this theoretical cross-fertilization seems to have emerged quite quickly. Although, as Jeppe Jensen (2009) also notes, the origins of CSR are usually dated to the publication of Thomas Lawson and Robert McCauley's (1990) *Rethinking Religion: Connecting Cognition and Culture*, some of the basic premises of CSR were presented by Stewart Guthrie in his essay *A Cognitive Theory of Religion*, published a decade earlier in 1980, in which he ties religion tightly to the universal human tendency to interpret ambiguous phenomena in anthropomorphic terms. This essay is the foundation for his subsequent book *Faces in the Clouds: A New Theory of Religion*, published in 1993, in which he develops his argument more clearly in evolutionary terms, and suggests that the tendency both to animate and anthropomorphize the world arises from certain evolved cognitive predispositions, which must have conferred huge adaptive advantages upon early humanity.

Evolutionary theory may not, then, have been a focal point of the very earliest research in CSR (as Pyysiäinen (this volume) also observes), but once the connection was made between certain aspects of religion and certain, seemingly natural, cognitive predispositions, the emergence of a specifically evolutionary cognitive science of religion was surely inevitable. However, this union between various evolutionary theories and certain aspects of CSR is not universally supported. Paul Thagard (2005), for example, is not unsympathetic towards CSR in general, but in reviewing some of the ways 'in which evolutionary psychology might be applied to explaining religious belief and practice', he finds 'no good reason to apply any of them' (p. 69). Whitehouse, on the other hand, in an essay describing future directions for the field, argues that the study of cognition may enhance our understanding of religious evolution whilst acknowledging it is too early '... to say much with confidence about the relationship between cognitive evolution and religion because of the paucity of evidence in evolutionary psychology generally and in its claims about religious phenomena in particular' (2008, pp. 32–3). Like Barrett, Whitehouse's own view of CSR seems to be that it does not strictly need to be supplemented by evolutionary psychology to remain a credible and enlightening enterprise, even if he recognizes the potential significance of certain evolutionary claims. Others, including Robert McCauley (2011), are relatively sanguine about the utility of evolutionary models in the field, but seem careful not to offer them unqualified support. If nothing else, McCauley argues, evolutionary psychological accounts, '... inspire provocative, new models about the modern human mind that have consequences that are experimentally testable and that connect directly with long-standing work in cognitive psychology' (p. 57). This is the case, McCauley argues, even if the data such models depend upon can be interpreted in alternative ways. As such, he implies, we should take their implications for the study of human life, including religion, very seriously.

As well as Guthrie, Steven Mithen (1996), Scott Atran (2002), Atran and Henrich (2010), Pascal Boyer (2002, 2003, 2010), Boyer and Bergstrom (2008), Jesse Bering (2006, 2011), Daniel Dennett (2006), Edward Slingerland (2008), Scott Tremlin (2006) and many others have all explicitly appealed to evolutionary theory of one kind or another in explicating the role played by universal features of cognition in constraining aspects of religious belief and behaviour. From the perspective of ECSR, many of religion's apparently recurrent features—including, for example, belief in supernatural agents and belief in the efficacy of certain kinds of ritual—are rooted firmly in our uniquely human evolutionary history. As Boyer writes, in discussing the progress and prospects of ECSR, 'The way animate beings are detected, agents represented, moral intuitions processed or contagion feared are all plausible outcomes of evolutionary processes. There is now a growing body of evolutionary thinking that connects the following elements of a potential evolutionary framework: (1) features of religious concepts; (2) experimental evidence for underlying cognitive systems; (3) clues about the genetic basis of these systems; and (4) precise hypotheses about the reproductive advantage provided by possession of such capacities' (2003, p. 123).

Certainly, evolutionary themes have already been explored in pursuit of a wide range of different objectives, from correlating accounts of cognitive evolution with archaeological evidence of quasi-religious practices and beliefs, to justifying the role of a reductive naturalistic epistemology in religious studies, and explaining the ultimate origins of belief in gods. Unsurprisingly, given the breadth of the field, there is considerable disagreement between different theorists who nonetheless each want to advance evolutionary arguments. Some of these internal disputes appear relatively minor, such as the debate between those, including Bering, Dennett, and Guthrie, who believe that the evolution of HADD plays a very significant role in the historical emergence of religion, and those such as Boyer who don't. Others, on the surface at least, appear to be of greater consequence. Possibly the most significant of these debates concerns the question of religion's adaptiveness—a question that is tackled explicitly, and in great detail in many of this volume's chapters. Indeed, since this particular debate traverses so many others in ECSR, it might usefully serve as a lens here to focus our discussion of the field and help develop a number of interrelated observations.

From the early days of ECSR—perhaps, as Sosis (2009) notes, because the field's pioneers were cognitive scientists—the dominant position has been that religion is a by-product of evolved cognitive mechanisms, rather than an 'adaptive complex of traits incorporating cognitive, affective, behavioral, and developmental elements' (p. 316). This debate should not be confused with the debate about the value of evolutionary theory in explaining religion per se. Rather, Boyer and Bergstrom (2008, p. 115) argue, it is about whether religious beliefs and behaviours are themselves adaptations: 'The first strategy tries to proceed directly from religious behavior to hypotheses about evolutionary processes that would have led to such phenomena... Alternatively, religious behavior may be explained in terms of underlying capacities or dispositions, which themselves are seen as outcomes of evolution'.

The adaptationist/by-product question is such a recurrent theme in the evolutionary study of religion generally, that particular evolutionary accounts are often categorized according to which side of this fence they sit, rather than by their more specific objectives, theoretical foci, or the data sets with which they are concerned. Unfortunately, such a simplistic differentiation of evolutionary theories carries a risk of occluding other interesting cases of theoretical divergence, and encourages the perception that ECSR, which is sometimes taken to be synonymous with the by-product thesis, is more unified than it really is. Sosis is quite right to claim that the adaptationist/by-product debate 'is likely to hamper efforts at genuine interdisciplinary work between evolutionary scientists and religious scholars because it reveals disagreements about the core ideas upon which the evolutionary study of religion is founded' (2009, p. 316), but we should perhaps be cautious regarding the nature and extent of ECSR's core. It may actually be rather smaller than is commonly recognized. Regardless, the adaptationist/by-product debate reveals at least three things about ECSR of great relevance to any attempt to assess the overall prospects and coherence of the field.

First, it clearly demonstrates the extent to which ECSR is characterized by theoretical pluralism. Yes, the dominant position is that religion is in some sense a by-product of cognitive evolution, but there is still quite a broad spectrum of opinion. ECSR is certainly not synonymous with any *particular account* of religion as a cognitive by-product. Paul Thurow (2011) helpfully divides ECSR into three camps—those who believe religion is a by-product of evolved cognitive mechanisms, those who believe religion is adaptive, and those who believe religion is a by-product of evolved cognitive mechanisms that subsequently became adaptive. As Peter Richerson and Leslie Newson (2009) make clear in their provocative essay, *Is Religion Adaptive: Yes, No, Neutral. But Mostly We Don't Know*, the debate is far from settled. Indeed, as Jeffrey Schloss (2009) explains in his introduction to *The Believing Primate: Scientific, Philosophical and Theological Reflections on the Origin of Religion*, which he edited with Michael Murray, we should not even take it for granted that the question itself is coherent, given the range of possible meanings of the terms 'religion' and 'adaptation' that are deployed in the literature.

Secondly, the debate between adaptationists and supporters of by-product hypotheses exposes the complex relationships between ECSR and other kinds of evolutionary approach to religion. After all, if ECSR is not necessarily committed to one or other of these perspectives, we cannot assume that attitudes towards the adaptationist/by-product debate are what primarily differentiate ECSR from other sorts of evolutionary approach. Evolutionary accounts of religion, needless to say, take a number of quite different forms, not all of which are interested in explaining religion exclusively, or even primarily, at the cognitive level. Amongst these approaches are commitment signalling theory (e.g. Bulbulia and Sosis 2011), cultural evolutionary approaches (see Wilson 2003, Richerson and Newson 2009 and this volume, Bainbridge (this volume)), religion as a meme (e.g. Dennett 2006), so-called neurotheological approaches (e.g. Newberg, D'Aquili, and Rause 2002), and religion as a system for regulating human interaction with the environment (e.g. Dow 2006). Contributing still further to the overall complexity and theoretical pluralism of this field is the fact that different surveyors of the field tend to divide the various approaches into quite different categories. Joseph Bulbulia (2004), Todd Tremlin (2012) and Michael Ruse (this volume) have all neatly summarized a variety of different extant approaches to the evolution of religion, but they have gone about it in different ways and reached quite different conclusions about which areas of research have the greatest prospects.

Most, in fact, are now inclined to accept that religion is a complex multi-faceted phenomenon, the evolutionary history of which cannot adequately be explored through the simple idea that religion is *either* a by-product *or* adaptive. Arguably, in recent years, there has been greater movement towards the integration of by-product accounts with other evolutionary approaches that do place a greater emphasis upon the adaptiveness of religion. Atran and Henrich, for example, appear to be seeking precisely this sort of grand integration when, in the introduction to their article, they reveal their intentions to argue '... that the cultural evolution of prosocial religions and the historical rise of large-scale civilizations involve the dynamic interaction of the by-products of adaptive cognitive mechanisms (e.g. minimally counterintuitive beliefs and overextended agent concepts), adaptive learning heuristics (e.g.

emulation of successful and prestigious individuals), credibility-enhancing ritual displays (e.g. self-sacrifice and costly commitments to seemingly preposterous beliefs), and cultural group selection for those packages of rituals, devotions, and beliefs that best sustain in-group prosocial norms (e.g. monumental undertakings, sacred values)' (2010, p. 19).

Thirdly, the adaptationist/by-product debate reveals a striking further difference between two different explanatory goals of ECSR—some accounts are primarily concerned with explaining the ultimate origins of religion, and others are mainly interested in explaining the spread, transformation and persistence of religion. The former sort of account typically draws upon evolutionary psychological ideas in explaining the development of a particular cognitive architecture, which gives rise to the distinctive mental tool-kit that makes the emergence of proto-religious ideas almost inevitable. For Boyer, this is the moment 'when such representations could occur in people's minds and exert enough fascination to be painstakingly translated into material symbols' (2002, p. 373). The latter, on the other hand, although it often shares the former's interest in cognitive architecture, is more concerned with explaining how particular features of distinctively religious beliefs and behaviours interact with particular features of the human cognitive system, in order to account for certain remarkable features of religions past and present (including their resilience, their universal preoccupation with supernatural agents, the efficacy of certain rituals, and so on). Although by-product theories have made important contributions to this second sort of account (see, for example, Atran (2002), Barrett (2004), Boyer (2002, 2010)), many have argued that the spread and transformation of religion over time cannot be explained adequately without understanding its adaptiveness (see, for example, Bering 2011; Wilson 2003). Clearly, the two approaches might be integrated to an extent (as attempted by Atran and Henrich, for example), but they remain distinctive sorts of enterprise nonetheless. Thurow (2011) even suggests that their relative strengths in dealing with the emergence and subsequent development of religion, respectively, both distinguishes by-product from adaptationist theories and demonstrates the need for a combinatorial approach. The former, he suggests, are best seen as theories of the origins of religious beliefs in normal cognitive processes in normal environments. They also explain why religious beliefs are widespread and enduring. By contrast, he argues, adaptationist theories explain why particular religious beliefs become so central to human communities.

Even this brief summary is sufficient to demonstrate the many subtle ways in which the central debates of evolutionary studies of religion are interrelated, and the many specific theories, commitments and objectives that must be disentangled in reflecting upon the prospects of ECSR as a whole. Keeping its inherent complexity and theoretical pluralism in mind, we turn briefly now to the question of how ECSR has been received in the broader field of religious studies.

EVALUATING ECSR

Criticisms of CSR and ECSR have come from many different angles, but we might discern three broad categories that are common to both fields. First, there are those who have highlighted technical difficulties with the psychological methods and models that underpin CSR and ECSR. A number of otherwise broadly sympathetic psychologists have addressed perceived methodological problems, and articulated the need for wider, deeper empirical study of the cognitive foundations of religious belief and practices, particularly the role played by affective and other bodily processes (see, for example, Gibson 2008; Sjöblom 2008). Other critics have raised questions about the information-processing model of mind it typically employs and its modular accounts of cognitive architecture. Nathaniel Barrett (2010), for example, has argued that the narrow understanding of computationalism that characterizes CSR is now widely discredited, even amongst cognitive scientists. Herrnstein Smith (2009), James Van Slyke (2011), Matthew Day (2004), and others, have all drawn attention to the various deficiencies of exclusively modular concepts of mind, and still others have complained at great length about the idea that cognitive devices of any kind can be said to be truly 'natural', in the sense of being strictly genetically determined (e.g. Karmiloff-Smith 2000).

Regarding the need for CSR to develop theoretically and the need for further and more detailed experimental data (in order to understand more fully the ways in which ToM or HADD, for example, interact with other psychological structures and processes), most CSR theorists are happy to agree. This can be seen as part of the natural evolution of the field. But criticisms of the underlying model of mind are contested rather more sharply. Here, responses to critics range from simple reaffirmations of mental modularity (it is simply the best available explanation of extant empirical evidence), to the attempt to establish compatibility between different models of cognitive architecture. Most, however, seem simply to have distanced themselves from the more controversial claims of the massive modularity thesis embraced by early evolutionary psychologists (see Buss 2005; Cosmides and Tooby 1992, 1994), according to which mental modules are, amongst other things, domain specific, innate, largely automatic systems, developed independently of social learning or particular cultural stimuli. Justin Barrett, as we have already observed, avoids talk of mental modules altogether, offering instead functional definitions of cognitive 'devices' or 'tool-kits'. And both he and McCauley prefer to talk about the maturational naturalness of 'features of cognitive systems' (McCauley 2011, p. 59) rather than 'innateness'.

Secondly, a number of critics, particularly philosophers, have addressed the implicit, and occasionally explicit, anti-religious tone adopted by some CSR (and many ECSR) theorists. The main concern here has been with the argument that CSR shows religion (specifically, belief in God) to be the result of cognitive malfunctioning, *and nothing more* (see, for example, Plantinga 2009; Leech and Visala 2011). It should already be clear that this is not a position with which many actually working in the field would agree. It is interesting to note in this context, however, that CSR's most conspicuous claim—the claim that religion is somehow 'natural' (unlike atheism)—has happily been embraced both by some religious believers and by some prominent atheists, who each interpret it in different ways. The latter group typically assumes that this undermines the intellectual case for religion, as it is portrayed more as the result of an unconscious perceptual predisposition than a well-reasoned set of beliefs or worldview. It is taken to be another victory in a long line of victories for science over religion. Many religious believers, on the other hand, take CSR's demonstration of the apparent naturalness of religion to be utterly unsurprising. Justin Barrett, for example, is adamant that his own cognitive explanation of why people naturally believe in gods does not detract from the truth of those beliefs in the slightest. Both positions have been debated at length.

A third general category of criticism comes mainly from those social and cultural anthropologists who are suspicious of the attempt to explain complex sociocultural phenomena such as religion in the terms, theories and concepts of biology, psychology and the cognitive sciences. James Laidlaw (2007), for example, has drawn attention to the narrowness of the concept of religion that he assumes underpins the whole of CSR, and its limitations from the perspective of the social sciences and other branches of religious studies. This criticism seems particularly stinging for those CSR theorists (including Atran and Boyer) who have approached the field first and foremost as anthropologists seeking to bring ethnographic research into contact with naturalistic scientific approaches to religion. From Laidlaw's perspective, the reduction of the study of religion to the answering of questions about the origins of religious beliefs and the cognitive foundations of particular rituals, contrasts so strongly with humanistic approaches to religion that any conversation between the fields will be extremely limited, even if they might be seen as compatible in a very loose sense (cf. Saler 2004).

In response to this line of criticism, however, it is frequently claimed that, with some possible exceptions (e.g. Slingerland 2008), the aim of those working in CSR is not primarily to reproach or marginalize other accounts of religion, such as those advanced by social and cultural anthropologists, sociologists, or other sorts of psychologist. Nor do they wish to replace the established methodologies of religious studies with something closer to the methods of experimental psychology or the cognitive neurosciences. In drawing attention to the fundamental roles played by universal cognitive processes in shaping religious beliefs and behaviour, CSR clearly challenges some central principles of certain strands of religious studies, but it constantly battles against the unjust criticism that it is intent on reducing the study of religion entirely to the study of cognitive processes. CSR does not deny the complex sociocultural heritage of particular religious ideas and behaviours, and has taken great care to emphasize the 'cultural scaffolding' of particular religious ideas by other complex, socially acquired knowledge, even if they are 'anchored' in the cognitive system. As Paul Bloom notes, 'Nobody is born with the idea that the birthplace of humanity was the Garden of Eden, or that the soul enters the body at the moment of conception' (2009, p. 120). Some theorists, notably David Sloan Wilson, have even explored the possibility of bridging the gap between humanistic/social-scientific and naturalistic approaches to religion in ways that recognize the significance of the social constructionist critique of CSR (see, for example, Wilson 2009).

Each of these criticisms of CSR might also be levelled at ECSR, but ECSR's alliance with evolutionary theory attracts a range of additional criticisms. This is despite the fact that most of its critics are very sympathetic to evolutionary biology, and some are also sympathetic towards many of the more moderate claims of sociobiologists. This is, perhaps, inevitable given that ECSR, like all evolutionary approaches to the study of religion, draws upon theories, concepts and methods that many within the natural and human sciences deem to be of questionable value in the explanation of sociocultural phenomena. As with CSR, then, dissatisfaction with ECSR is not the sole preserve of religious apologists or theologians who perceive an inherent marginalization of theological explanations for the origins of religious belief.

In particular, the alliance that some CSR theorists have forged with evolutionary psychology (EP)—which is itself the subject of many well-known critiques (as detailed in, for example, Rose and Rose 2000; Richardson 2007)—has drawn substantial, frequently adverse attention. In this context, serious doubts have been raised about EP's neglect of socioculturally contingent factors affecting the historical development of religion, its underlying methodological individualism, and the relative paucity of reliable evidence in the fossil record (or anywhere else for that matter), as well as questions about the reductive functional definitions of religion offered by some theorists, and the implicit atheistic agenda of others. Some of these criticisms, especially those relating to inadequate functional definitions of religion (although the problem is clearly not unique to this field), can also be applied to other evolutionary approaches to religion, but at least some manifestations of ECSR have appeared more vulnerable to the charge of being unnecessarily narrow and reductive. Notions of modularity and innateness are, inevitably, recurrent concerns, as is exasperation at the failure of some ECSR theorists to move beyond the notion that religion is a by-product of cognitive evolution, with no adaptive value of its own. Even some who happily acknowledge the intrinsic value of CSR in the study of religion (including Sosis (2009), Wilson (2009)) seem to find the refusal of some to accept the value of other evolutionary accounts of religion, including group selection and cultural evolutionary approaches, perplexing.

Arguably, a disproportionately high volume of the criticism received by both CSR and ECSR has been rather general and focused upon the grand, headline-making claims and the more controversial of the theories that underpin them. This may seem fair enough given the provocative titles of some of the more popular books to have made use of the work of cognitive scientists: 'Why Would Anyone Believe in God?'; 'The God Instinct'; 'Religion Explained'. As usual, the less attention-grabbing, often more technical, yet equally significant background details, which constitute the bulk of the actual research, are the last to receive sustained consideration from critics. General criticism, most frequently aimed at the evolutionary psychological commitments of some ECSR, has been politely received, and often relatively succinctly addressed through the qualification or moderation of certain claims where required. We have already noted that talk of cognitive tool-kits or devices has largely replaced talk of mental modules in ECSR, and other helpful clarifications abound. It is now frequently emphasized, for example, that CSR research explains the likelihood of the emergence and persistence of religious belief in historical and contemporary societies rather than its inevitability, and that religion is not just an evolutionary by-product, nor simply adaptive. ECSR's supporters have also sometimes argued that it is not, strictly speaking, religion per se they are trying to explain. Rather, ECSR explains quite specific things like the origins and persistence of certain aspects of religion, or the tendency to infer the activity of supernatural agents in the world, or why we find gods so easy to believe in. Other issues relating to religion are of much less interest to ECSR, including explaining the particular forms of religious beliefs and behaviours. This process of clarification, qualification, and moderation of some of the more controversial claims and theories has certainly countered much of the more prominent criticism levelled at the field, though the extent to which these issues have been fully resolved (as many of the essays in this book argue) remains open to question.

Despite the existence of a few, key, widely endorsed theories and notions in ECSR, it would clearly be a mistake for its critics to tar the whole field with the same brush. The emergence of a 'standard model' may create the illusion of universal consensus, but, it is worth reiterating, both CSR and ECSR are very complex and diverse fields, which are constantly evolving. ECSR is not just unreconstructed evolutionary psychology enlisted in the study of religion. And nor should it always be seen as an attempt to reduce religion to biology, nor an assault on theism itself. Of course, particular ECSR theories will always be vulnerable to criticism of some kind (it would be very curious indeed if they were not), but they are not all equally vulnerable to the same criticisms. There are relatively few aspects of ECSR that are universally agreed upon, and many different theories and ideas have been recombined in any number of different ways in the explanation of various aspects of religion. Critics, then, would be well advised to choose their targets carefully, and resist the temptation to condemn the whole field as a result of dissatisfaction with a particular theory or theorist. Indeed, uncompromisingly condemnatory attitudes to the field in general, especially from those who wish to preserve the multidimensionality of religious studies, are surely misconceived. The study of the evolutionary cognitive foundations of the capacities for various sorts of religious activities is not necessarily any less intrinsically valuable than any other area of the study of religion, even if it is unlikely to provide all-inclusive explanations of the phenomena with which it is concerned. Fortunately, very few of ECSR's critics even come close to calling for the field to be abandoned in its entirety, even if some apparently believe fundamental changes are necessary to some core (if not quite universal) principles.

More targeted criticism, on the other hand, will encourage ECSR's supporters to clarify, elaborate, and perhaps even to moderate still further any less convincing or more overstated claims. Careful consideration of how ECSR might be further expanded and improved as a result of deeper engagement with ethnographic research, other branches of religious studies, other psychologies, and other evolutionary accounts of religion is also essential. ECSR's roots are proudly multi-disciplinary, and interdisciplinary dialogue must continue if it is to be more fully integrated into religious studies, as many of its supporters hope it will be. Even if establishing ECSR's compatibility with other disciplines may seem a relatively low hurdle, genuinely constructive and coherent dialogue, in which ECSR recognizes its own theoretical and practical limitations in various areas, as well as its unique strengths, is a more ambitious (and theoretically fraught) project. Certainly, there has been great progress in this direction in recent years, but there remains a great deal of work that could usefully be done.

THE AIMS OF THIS BOOK

As its title suggests, this book has two primary aims. First, it aims to ask critical questions of ECSR from anthropological, philosophical, psychological, sociological, cultural evolutionary, and theological perspectives. But as well as providing a critical counterbalance to the enormous weight of recent publications that are broadly supportive of ECSR, this book also offers a number of constructive suggestions regarding the future development of the field. None of the contributors believe that ECSR has no place in the study of religion, but all believe it faces a number of theoretical, evidential, and methodological challenges, which need to be addressed if the field is to advance. The balance between critique and constructive proposals varies chapter by chapter, but all might be seen as participating in the development rather than the destruction of ECSR, where a commitment to the continuous testing and evaluation of evidence, and the updating and improvement of models and theories, remains central to its scientific credentials.

Inevitably, a number of compromises have had to be made in compiling this volume. First, and most importantly perhaps, the decision to invite a smaller number of longer, more detailed essays means this book cannot claim to represent the full range of technical debates within ECSR. This is a consequence of our desire to emphasize the multi-disciplinary context of ECSR. We have tried to offer a representative array of detailed discussions, and produce a valuable and accessible resource for researchers and students from a broad spectrum of academic fields beyond the established ECSR community. Furthermore, this volume does not seek to provide a comprehensive overview of all evolutionary theories of religion more generally. Most of the contributors are deeply concerned at some level with the relationship between ECSR and other evolutionary accounts, and some even argue that other sorts of evolutionary account provide a crucial supplement to ECSR, but it has not been possible to cover the evolution of religion as thoroughly as we would have liked. Nevertheless, the two opening chapters, by Ilkka Pyysiäinen and Michael Ruse respectively, do attempt to set ECSR in broader theoretical contexts. In outlining the scope and central themes of CSR and ECSR Pyysiäinen offers a brief historical survey of the fields, which makes the breadth of their theoretical influences, and the extent of the data upon which they draw abundantly clear. Michael Ruse, by contrast, presents a helpful introduction to some of the key theories and debates in the evolutionary study of religion from the perspective of evolutionary biology, and in so doing shows how ECSR is related to much wider debates about the evolution of culture generally, and religion specifically.

To many CSR and ECSR theorists, it may seem strange that a book in this field does not list amongst its primary aims the presentation of new empirical evidence supporting or confounding ECSR. Undoubtedly, most contributors are primarily concerned to offer detailed conceptual analysis of various aspects of ECSR—a vitally important enterprise, which cannot be considered of secondary importance to empirical research. But that is not to say that the book as a whole is not deeply concerned with the evaluation of empirical evidence. The various essays differ markedly, however, in the specific use that they make of empirical data and studies. Some are critical of the conclusions drawn from existing empirical studies, and point towards alternative conclusions and the implications for particular evolutionary theories of religion, whereas others introduce novel empirical studies to the subject. Fraser Watts' essay, for example, draws upon contemporary neuroscientific studies (including studies of Interactive Cognitive Subsystems and Ian McGilchrist's recent work on hemispheric lateralization) in developing his argument that those accounts of cognitive evolution currently dominating ECSR are not the only plausible candidates. Purzycki, Haque, and Sosis review a considerable range of empirical studies related to the themes of commitment signalling theory, and supernatural punishment and reward in exploring the potential of a 'dynamic systems' approach to religion. Newson and Richerson, on the other hand, draw upon an impressive range of empirical studies of cultural transformation to present an original cultural evolutionary approach to understanding religious change at a societal level. Bill Bainbridge also offers empirical support for a cultural evolutionary approach to religion in explaining how work in the field of artificial intelligence (AI) can elucidate the spread of religious ideas. Although the absence of much *novel* empirical data in this book may (sadly) make it relatively less attractive to some ECSR researchers, we hope that the editorial decisions we have taken will enhance its appeal to others working in the humanities and social sciences seeking a critical introduction to many of the important debates surrounding cognitive scientific approaches to religion.

A number of subjects are addressed by several different authors, but two are especially prominent. First, many of the chapters tackle questions relating to evolutionary psychology in considerable detail. Chapters by Purzycki, Haque and Sosis, Barrett, Watts, Jenkins, Newson and Richerson, Bainbridge, and Turner seek to do much more than simply criticize ECSR for its affiliation with what many outside the field see as a deeply problematic approach to understanding human cognition and behaviour. All are concerned with isolating very particular aspects of evolutionary psychology that raise problems for the study of religion, and identifying alternative ways in which the evolutionary study of religion might develop. Nevertheless, some tackle this issue more directly than others. The chapters by Tim Jenkins, Léon Turner, and Aku Visala, in particular, aim to explore the relationship between ECSR and humanistic/social-scientific accounts of religion (in Jenkins' and Turner's case), and between ECSR and theism (in Visala's case). Each questions widely held beliefs about intertheoretical compatibility, and each identifies the most plausible strategies for pursuing future interdisciplinary research. Through an original comparative analysis of two studies of witchcraft, Jenkins' critique of ECSR highlights differences between evolutionary cognitive and anthropological notions of the relationship between history and nature. Ultimately, he concludes, reconciliation between naturalistic and historical accounts of religion should be brought about through the historicization of the notion of evolutionary processes. Turner takes CSR's claims about belief in gods as his case study. He seeks to identify and differentiate key claims and theoretical commitments of CSR and ECSR, and to explain precisely what, if anything, might lead to conflict with sociohistorical approaches to religion. Turner finally concludes that CSR will need to abandon the naturalistic causal explanations of belief in gods that characterize ECSR if it is to preserve the benefits of its relationship with humanistic and social-scientific studies of religion. Visala also differentiates CSR from ECSR and seeks to identify specific areas of potential conflict between them and theism. His subtle and innovative philosophical analysis leads him to conclude that ECSR raises no more problems for theism than evolutionary biology.

A second recurrent theme concerns the way various aspects of religion are shaped by the interaction between individual human beings and their social, cultural and physical environments. Again, some contributors tackle this issue more directly than others. Bill Bainbridge's AI approach, for example, emphasizes the social nature of religious cognition—how it is embodied in social networks and groups—and seeks to explain how distinctive religious movements may expand or contract within particular communities as a result of competition between them. Bainbridge studies the distribution and, ultimately, the success of hypothetical religious movements through modelling possible processes affecting the formation and dissolution of social bonds between immediate neighbours in a social network. Nathaniel Barrett, by contrast, brings ECSR into contact with the work of the philosopher Robert Cummings Neville and the ecological psychologist James Gibson. Concentrating on the notions of value and engagement, Barrett departs radically from ECSR's account of the production of religion through the mechanical interaction of natural cognitive devices and physical and social environments. He explains how both environments and our experiences of them might be seen as value rich and seeks to reconceptualize interactive cognition as 'engagement with value'. Purzycki et al. emphasize the socioecological contexts of religion throughout their essay, describing a variety of social functions of different religious behaviours whilst demonstrating the explanatory benefits of an adaptationist account of religion that complements CSR's better known theses concerning religious cognition. Via a discussion of the functional logic of the constituent parts of religious systems, they offer novel suggestions as to how such dynamic systems might adapt to environmental change. Approaching the evolution of religion from a completely different direction, van Huyssteen's main concern is with human embodiment. From this perspective, bodily processes are essential components of all aspects of human life, and an understanding of how these processes shape our experiences of ourselves and our world is central to understanding our evolutionary history. Noting ECSR's focus upon the evolution of specialized cognitive tools and its neglect of bodily constraints upon cognition, van Huyssteen insists on supplementing our evolutionary understanding of religion with a broader embodied account of what it means to be a human person.

Given the particular subject of this volume, a few final comments might usefully be made about its contributors. Despite the fact that many of us (including the editors) are employed in theology and religious studies departments, none of the essays in this book seek to promote any particular religious or theological agenda. Van Huyssteen, Visala, and Watts are all concerned in different ways with the claim that religion is somehow a cognitive error, and that ECSR has somehow exposed the falsity of religious beliefs. But none of them explicitly argue for the opposite position, here—that ECSR (or any other account of the evolution of religion for that matter) might demonstrate the truth of religion. The contributors are also united in other ways. For example, all are agreed that Darwin's theory of evolution by natural selection is amongst the most successful biological theories of all time, and none reject the possibility that evolutionary theory has something important to contribute to the study of religion, which, it is agreed, is an inherently interdisciplinary pursuit. All also share the belief that the study of religion cannot be reduced to the study of psychological structures and processes. In short, we believe the contributors to be broadly representative of those with an interest in ECSR, the vast majority of whom have no particular axe to grind with either scientific or religious accounts of religion.

REFERENCES

Atran, S. 1993. Whither 'ethnoscience'? In *Cognitive Aspects of Religious Symbolism*. Ed. Pascal Boyer. Cambridge: Cambridge University Press, 48–70.

——. 2002. *In Gods We Trust: The Evolutionary Landscape of Religion*. Oxford: Oxford University Press.

Atran, S. and J. Henrich. 2010. The evolution of religion: How cognitive by-products, adaptive learning heuristics, ritual displays, and group competition generate deep commitments to prosocial religions. *Biological Theory 5*(1): 18–30.

Barrett, J. 2004. *Why Would Anyone Believe in God?* Lanham: Alta Mira Press.

Barrett, N. F. 2010. Toward an alternative evolutionary theory of religion: Looking past computational evolutionary psychology to a wider field of possibilities. *Journal of the American Academy of Religion 78*(3): 583–621.

Baumard, N. and C. Chevallier. 2012. What goes around comes around: The evolutionary roots of the belief in immanent justice. *Journal of Cognition and Culture 12*(1–2): 1–2.

Bering, J. 2006. The folk psychology of souls. *Behavioral and Brain Sciences 29*: 453–98.

——. 2011. *The God Instinct: The Psychology of Souls, Destiny, and the Meaning of Life*. London: Nicholas Brealey.

Bloom, P. 2009. Religious belief as an evolutionary accident. In *The Believing Primate: Scientific, Philosophical, and Theological Reflections on the Origin of Religion*. Eds. M. Murray and J. Schloss. Oxford: Oxford University Press, 118–27.

Boyer, P. 2002. *Religion Explained: The Human Instincts that Fashion Gods, Spirits and Ancestors*. London: Vintage.

——. 2003. Religious thought and behaviour as by-products of brain function. *Trends in Cognitive Sciences 7*(3): 119–24.

——. 2010. Why evolved cognition matters to understanding cultural cognitive variations. *Interdisciplinary Science Reviews 35*(3–4): 376–86.

Boyer, P. and B. Bergstrom. 2008. Evolutionary perspectives on religion. *Annual Review of Anthropology 37*: 111–30.

Bulbulia, J. 2004. The cognitive and evolutionary psychology of religion. *Biology and Philosophy 19*: 655–86.

Bulbulia, J. and R. Sosis. 2011. Signalling theory and the evolution of religious cooperation. *Religion 41*(3): 363–88.

Buss, D. M. Ed. 2005. *Handbook of Evolutionary Psychology*. Hoboken, NJ: Wiley

Cosmides, L. and J. Tooby. 1992. Cognitive adaptations for social exchange. In *The Adapted Mind*. Eds. J. H. Barkow, L. Cosmides, and J. Tooby. New York: Oxford University Press, 163–228.

——. 1994. Origins of domain specificity: the evolution of functional organization. In *Mapping the Mind*. Eds. L. A. Hirschfeld, and S. A. Gelman. Cambridge: Cambridge University Press, 85–116.

Day, M. 2004. The ins and outs of religious cognition. *Method and Theory in the Study of Religion 16*: 241–55.

Dennett, D. C. 2006. *Breaking the Spell: Religion as a Natural Phenomenon* (No. 14). New York: Viking Press.

Dow, J. W. 2006. The evolution of religion: three anthropological approaches. *Method & Theory in the Study of Religion 18*(1): 67–91.

Gibson, N. 2008. Once more, with feelings: The importance of emotion for cognitive science of religion. In *The Evolution of Religion: Studies, Theories, and Critiques*. Eds. J. Bulbulia, R. Sosis, E. Harris, R. Genet, C. Genet, and K. Wyman. Santa Margarita, CA: Collins Foundation Press, 271–8.

Guthrie S. E. 1993. *Faces in the Clouds. A New Theory of Religion*. New York: Oxford University Press.

Guthrie, S. E., et al. 1980. A cognitive theory of religion [and comments and reply]. *Current Anthropology 21*(2): 181–203.

Jensen, J. S. 2009. Religion as the unintended product of brain functions in the 'standard cognitive science of religion model': On Pascal Boyer, *Religion Explained* (2001) and Ilkka Pyysiäinen, *How Religion Works* (2003). In *Contemporary Theories of Religion: A Critical Companion*. Ed. M. Strausberg. New York: Routledge, 129–55.

Karmiloff-Smith, A. 2000. Why babies' brains are not Swiss army knives. In *Alas Poor Darwin: Arguments Against Evolutionary Psychology*. Eds. H. Rose and S. Rose. London: Jonathan Cape, 144–56.

Laidlaw, J. 2007. A well-disposed social anthropologist's problems with the 'cognitive science of religion'. In *Religion, Anthropology and Cognitive Science*. Eds. H. Whitehouse and J. Laidlaw. Durham, NC: Carolina Press, 211–46.

Lawson, E. T. and R. N. McCauley. 1990. *Rethinking Religion: Connecting Cognition and Culture*. Cambridge: Cambridge University Press.

Leech, D. and A. Visala. 2011. The cognitive science of religion: Implications for theism? *Zygon 46*(1): 47–64.

McCauley, R. N. 2011. *Why Religion is Natural and Science is Not*. New York: Oxford University Press.

McCauley, R. N. and H. Whitehouse. 2005. Introduction: New frontiers in the cognitive science of religion. *Journal of Cognition and Culture 5*(1–2): 1–12.

Mithen S. J. 1996. *The Prehistory of the Mind*. London: Thames and Hudson.

Murray, M. J. and A. Goldberg. 2009. Evolutionary accounts of religion: Explaining and explaining away. In *The Believing Primate: Scientific, Philosophical, and Theological Reflections on the Origin of Religion*. Eds. M. Murray and J. Schloss. Oxford: Oxford University Press, 179–99.

Newberg, A., E. G. D'Aquili, and V. Rause. 2002. *Why God Won't Go Away: Brain Science and the Biology of Belief*. New York: Ballantine Books.

Plantinga, A. 2009. Games scientists play. In *The Believing Primate: Scientific, Philosophical, and Theological Reflections on the Origin of Religion*. Eds. M. Murray and J. Schloss. Oxford: Oxford University Press, 139–67.

Pyysiäinen, I. 2002. Introduction: Cognition and culture in the construction of religion. In *Current Approaches in the Cognitive Science of Religion*. Eds. I. Pyysiäinen and V. Anttonen. London: Continuum, 1–12.

Pyysiäinen, I. and M. Hauser. 2010. The origins of religion: evolved adaptation or by-product? *Trends in Cognitive Sciences 14*(3): 104–9.

Richardson, R. C. 2007. *Evolutionary Psychology as Maladapted Psychology*. Cambridge: MIT Press.

Richerson, P. and L. Newson. 2009. Is religion adaptive? Yes, no, neutral. But mostly we don't know. In *The Believing Primate: Scientific, Philosophical, and Theological Reflections on the Origin of Religion*. Eds. M. Murray and J. Schloss. Oxford: Oxford University Press, 100–17.

Rose, H. and S. Rose. Eds. 2000. *Alas Poor Darwin: Arguments against Evolutionary Psychology*. London: Jonathan Cape.

Saler, B. 2004. Towards a realistic and relevant 'science of religion'. *Method and Theory in the Study of Religion 16*: 205–33.

Schloss, J. 2009. Introduction: Evolutionary theories of religion. Science unfettered or naturalism run wild. In *The Believing Primate: Scientific, Philosophical, and Theological Reflections on the Origin of Religion*. Eds. M. Murray and J. Schloss. Oxford: Oxford University Press, 1–25.

Sjöblom, T. 2008. Narrativity, emotions, and the origins of religion. In *The Evolution of Religion: Studies, Theories, and Critiques*. Eds. J. Bulbulia, R. Sosis, E. Harris, R. Genet, C. Genet, and K. Wyman. Santa Margarita, CA: Collins Foundation Press, 279–86.

Slingerland, E. 2008. 'Who's afraid of reductionism? The study of religion in the age of cognitive science'. *Journal of the American Academy of Religion 76*(2): 375–411.

Slingerland, E. and J. Bulbulia. 2011. Introductory essay: Evolutionary science and the study of religion. *Religion 41*(3): 307–28.

Slone, J. 2004. *Theological Incorrectness: Why Religious People Believe what they Shouldn't*. Oxford: Oxford University Press.

Smith, B. Herrnstein. 2009. *Natural Reflections: Human Cognition at the Nexus of Science and Religion*. New Haven, CT: Yale University Press.

Sørensen, J. 2006. *A Cognitive Theory of Magic*. Walnut Creek: AltaMira Press.

Sosis, R. 2009. The adaptationist-byproduct debate on the evolution of religion: Five Misunderstandings of the adaptationist program. *Journal of Cognition and Culture 9*: 315–32.

Thagard, P. 2005. The emotional coherence of religion. *Journal of Cognition and Culture 5*(1–2): 58–74.

Thurow, J. C. 2011. Does cognitive science show belief in god to be irrational? The epistemic consequences of the cognitive science of religion. *International Journal for Philosophy of Religion*, 1–22. Available online at <http://link.springer.com/article/10.1007/s11153-011-9300-y>.

Tremlin, T. 2006. *Minds and Gods: The Cognitive Foundations of Religion.* New York: Oxford University Press.
——. 2012. The Origins of Religion. In *Science and the World's Religions, vol. 1*. Eds. P. McNamara and W. Wildman. Santa Barbara: Praeger, 3–38.
Van Slyke, J. 2011. *The Cognitive Science of Religion*. Farnham: Ashgate.
Whitehouse, H. 2004. *Modes of Religiosity: A Cognitive Theory of Religious Transmission.* Walnut Creek: AltaMira Press.
——. 2008. Cognitive evolution and religion; Cognition and religious evolution. In *The Evolution of Religion: Studies, Theories, and Critiques*. Eds. J. Bulbulia, R. Sosis, E. Harris, R. Genet, C. Genet, and K. Wyman. Santa Margarita, CA: Collins Foundation Press, 31–41.
Wilson, D. S. 2003. *Darwin's Cathedral: Evolution, Religion, and the Nature of Society.* Chicago: University of Chicago Press.
——. 2009. Evolutionary social constructivism: narrowing (but not yet bridging) the gap. In *The Believing Primate: Scientific, Philosophical, and Theological Reflections on the Origin of Religion*. Eds. M. Murray and J. Schloss. Oxford: Oxford University Press, 318–38.

2

The Cognitive Science of Religion

ILKKA PYYSIÄINEN

THE COGNITIVE RESEARCH PROGRAMME

The term 'Cognitive Science of Religion' refers to a cross-disciplinary research programme in the study of religion that emerged in the 1990s (see Barrett 2007; Pyysiäinen 2008; Tremlin 2006). It got its start with the publication of Lawson and McCauley's (1990) *Rethinking Religion* and Boyer's *Tradition as Truth and Communication* (1990) and *The Naturalness of Religious Ideas* (1994). A cognitive perspective had been introduced to anthropology even earlier, however. Sperber's *Rethinking Symbolism* (1995/1975) was a strong argument against symbolist anthropology and semiotic approaches to cultural and artistic symbolism; according to Sperber, these should be replaced by cognitive explanations of the mental mechanisms that make symbolic interpretation possible in the first place (see Sperber and Wilson 1988, pp. 9, 255).

In Sperber's view, so-called symbolic exegesis consists of searches in memory in order to find a relevant place for new information in the preexisting data base, not of finding a meaning. Symbolic systems are self-referential and thus do not carry any meaning (but cf. Lawson and McCauley 1990, pp. 146–55). Symbolism is best viewed as a cognitive mechanism that participates in the construction of knowledge as well as in the functioning of memory. It is not an instrument of social communication or a property of phenomena that could be considered apart from this mechanism (Sperber 1995, pp. xi–xii, 146–7).

The symbolic process has two aspects: a displacement of attention (focalization) and a search in memory (evocation). In focalization, attention shifts from the new information in question to the unfulfilled conceptual conditions. In evocation, the new information is reviewed and tested against the information in one's long-term memory. When the invocation of relevant background information fails, evocation begins. It is this evocation that is substituted for 'meaning' in this model: the meaning of a 'symbol' is the same as its evocative processing. Thus, for example, the 'meaning' of the image of the cross consists of all the individual evocative searches that people undertake for a relevant place in memory for this piece of information. The symbolic mechanism creates its own pathways in memory, and the symbolic process may become endless. Religious representations, for example, are never given a final definition; the symbolic exegesis rather is an endless process (Sperber 1995, pp. 119–23, 141–5). Anthropologists should best avoid participating in symbolic exegesis by their interpretative efforts and, instead, should try to explain what makes symbolic exegesis possible in the first place.

Ten years after his first book, Sperber published a paper on what he calls the epidemiology of beliefs or of representations (Sperber 1985). Such epidemiology focuses on how religious concepts and beliefs are spread within and across cultures and why some representations are more 'contagious' than some others, as it were. Certain kinds of beliefs and other mental representations seem to be 'attractors' that call for our attention because of a natural fit with panhuman mental structures (see Atran 1987; Claidière and Sperber 2007, 2010; Sperber 1996).

Boyer (1994) then applies this model to explore the factors that contribute to the differential spread of religious ideas. Not all information is culturally transmitted, because much of our knowledge and assumptions are based on implicit and intuitive inferences that spring from the mind's panhuman intuitive ontology. We tacitly assign entities into ontological categories and make intuitive inferences concerning the properties of those entities in question. If, for example, an entity can cry, it obviously belongs to the category of persons and thus has the standard properties of persons. If an entity is 'being fixed' it obviously is an artifact and does not have the properties of a person (Boyer 1994, pp. 91–124; Gopnik et al. 1999; Hirschfeld and Gelman 1994; Keil 1979, 1996; Sperber et al. 1995). Consequently, models of cultural transmission that neglect the role played by implicit cognition and the inferential nature of communication are incomplete (Boyer 1994, pp. 270–8). Cognitive and developmental psychology are important as they provide information about the nature and gradual development of the human cognitive architecture that canalizes the recurrent patterns of thought and behaviour (Atran 2002; Barrett 2004; Boyer 1994, 2001, 2003; Pyysiäinen 2001).

Lawson and McCauley (1990; McCauley and Lawson 2002), for their part, focus on how the structure of religious ritual action is mentally represented in the same ways as any action: somebody does something to someone (using some instrument). Gods can be conceived of as either active agents doing something to humans via rituals (e.g. give their blessing), or as patients of human action as in sacrifices, for example. In the first case we have 'special agent' rituals which are ideally not repeated for one and the same patient: people are baptized, married, and buried only once, although certain controversial exceptions do exist (in Catholicism, divorce and remarrying are grave theological problems, for example). Divine actions have 'superpermanent' effects.

'Special patient' rituals, for their part, can be repeated time and again because human action cannot establish anything once and for all. What is important in all religious rituals is that they are collective and are meant to bring about some change in the religious world. Thus, praying silently alone is not a ritual in Lawson and McCauley's sense of the term. It does not have any commonly accepted and recognizable effects in the religious world.

Currently, we can identify such areas of research as Barrett and colleagues' work on the mental representation of non-natural agent concepts, Bering's experiments on the folk psychology of souls, Guthrie's theory of religion as a form of systematic anthropomorphism, Whitehouse's modes theory of religious transmission, and Boyer and Liénard's hazard precaution theory of ritualized behaviour (Barrett 2004, 2007; Barrett et al. 2001; Bering 2006; Bering and Bjorklund 2004; Bering and Parker 2006; Boyer and Liénard 2006; Guthrie 1993; Liénard and Boyer 2006; Pyysiäinen 2009; Whitehouse 1995). The significance of cognitive theories and findings for theology also has received attention lately (Barrett 2011; Visala 2011).

Besides, the cognitive science of religion has been introduced to a wider audience, interpreted, and elaborated in a number of publications (Martin 2003; Pyysiäinen 2001, 2004b; Tremlin 2006) as well as applied in the history of religions (Beck 2006; Czachesz 2007; Luomanen, Pyysiäinen, and Uro 2007; Martin 2004; Pyysiäinen 2004c, 2009; see Pyysiäinen 2008). It thus has now grown into a wide research programme that draws not only from cognitive science but also from evolutionary and developmental psychology, neuroscience, computer science, ethology, and anthropology (see Barrett 2007; Pyysiäinen 2008, 2012; Pyysiäinen and Hauser 2010). The major defining characteristics of this programme are a strong commitment to methodological naturalism and the view that anthropology and sociology should not be insulated from psychology (Boyer 1994; Pyysiäinen 2001, 2012). Naturalism here means that explaining religion as human behaviour necessitates neither special 'religious' methods nor a personal commitment to beliefs about the supernatural (or their rejection). As for psychology, some cognitive scientists of religion seem to favour psychological reductionism while others emphasize that religion can be studied at differing levels and that the nature of questions asked determines at which level the answer is sought (see McCauley 2011; Pyysiäinen 2009, in press a).

Evolution was not a prominent topic in the beginnings of CSR (*pace* Boyer 1994, pp. 263–96). Later, Boyer spent time in Santa Barbara with Tooby and Cosmides, and evolutionary issues consequently play a more focal role in *Religion Explained* (Boyer 2001). At the same time other evolutionary considerations of religion as an adaptation emerged (e.g. Wilson 2002). Yet the evolutionary perspective is not a necessary element in CSR; Thagard (2005, pp, 69–72), for example, rejects it outright. Pyysiäinen (2006) takes up a novel point in considering the Baldwin effect (see Depew 2003): learned behaviours can affect the direction and rate of evolutionary change by natural selection because they enable individuals to modify the context of natural selection that affects their offspring who thus face new kinds of selection pressures. If there is a genetically determined 'good trick' that increases one's fitness, then in each generation individuals will have to be better and better at the trick to outcompete others until, finally, the ability to perform the trick will have to be innate (Dennett 1991, pp. 184–7). In the case of religion, this does not necessarily mean genetic assimilation as cultural change is more rapid than biological evolution; thus, for example, religion may not have shaped the human brain, but its evolution may rather have been guided by the fact that some forms of religion (those which were easier to adopt) have survived better than others (see McCauley 2011). Religion has not become genetically assimilated but, rather, has coevolved with the human brain (see Godfrey-Smith et al. 2003, p. 112; Pyysiäinen 2006, pp. 213–17).

However, recently religion has been studied as a biological adaptation or at least as adaptive in cultural evolution (see Boyer and Bergstrom 2008; Pyysiäinen and Hauser 2010; Schloss and Murray 2011; Wilson 2002). Religion has been regarded as a solution to the evolutionary puzzle of altruism and cooperation: why do individuals engage in altruistic behaviours even at the cost of their genetic fitness (Axelrod 1990; Hamilton 1964; Henrich and Henrich 2007; Trivers 2002)? Some argue that religion has directly contributed to the evolution of intra-group cooperation and others argue that it has forced individuals to refrain from cheating because of the fear of supernatural punishment (see Schloss and Murray 2011). In the first case, religion works through cultural group selection (see Richerson and Boyd 2005) while in the latter case the unit of selection is either the individual or a gene.

Bering argues in line with the latter alternative that there is a cognitive system dedicated to forming illusory representations of psychological immortality and symbolic meaning; this system has evolved as a response to the unique selection pressures of the human social environment. A 'representational bias for envisioning personal immortality' has enhanced the genetic fitness of individual humans in ancestral environments. Beliefs about ghosts and afterlife thus are illusory but adaptive beliefs (Bering 2006, 2011). Similarly, beliefs about an all-seeing god may make people act morally because even in the absence of other humans there still is god who sees and remembers our good and bad deeds and consequently punishes and rewards us (Bering and Johnson 2005). Atkinson and Bourrat (2011) found in their study of 87 countries that beliefs about God and afterlife, indeed, independently predict the assessment of the justifiability of a range of moral transgressions and that beliefs about the permissibility of moral transgressions are tied to beliefs about supernatural monitoring and punishment.

There is experimental evidence that even painting eyes over an honesty box at the office kitchen makes employees put their coins in the box more scrupulously (Bateson et al. 2006). But, as Schloss and Murray (2011, p. 56) point out, it seems to be enough that a person assumes that *someone* is watching; there is no need to postulate superhuman agency. Any kind of policing is enough. Johnson (2011), however, replies that a supernatural threat is far more efficacious than any natural counterpart. This requires that the person(s) in question really believe in the power and existence of supernatural agents, though. We may, then, ask which comes first, cooperation or beliefs about moralizing gods?

I tend to agree with Shariff (2011; see Sanderson and Roberts 2008) who argues that 'big gods' with omniscient scope and punitive abilities are relatively recent (5,000–10,000 years ago) innovations that have developed in large, complex societies. They are by-products of existing cognitive adaptations and have spread culturally rather than genetically. It is difficult to see religion as a biological adaptation for cooperation because its very existence requires such cognitive-emotional mechanisms that support prosociality and intragroup cooperation quite apart from any kind of religion. Notably, there is no special 'religion module' in the brain (Day 2009). Religion is more logically seen as a by-product of quite mundane, evolved cognitive mechanisms (Boyer 2001; Pyysiäinen 2012; Pyysiäinen and Hauser 2010).

REFLECTIVE THEOLOGIES AND INTUITIONS

Cognitive scientists have criticized anthropologists and scholars of religion for not having distinguished between theologies and cultural models, on the one hand, and people's actual beliefs and behaviours, on the other. Written theologies are not exhaustive catalogues of the beliefs of a given population. Instead, they are artifacts that serve as cues directing people's inferences; being exposed to some official theological formulations triggers mental representations that are more salient than others to start with. Although we often can predict more or less accurately what kinds of ideas people will end up entertaining after being exposed to certain written texts or sermons, there are no *necessary* links between these ideas and the theological stimuli or what Barrett (1999) calls 'theological correctness'. The actually represented ideas do not mechanically follow from the perceived stimuli; they are brought about by an active inferential or associative process in the mind of the person in question (Atran and Medin 2008; Sperber 2006; Sperber and Wilson 1986). There are no intrinsic meanings in written texts (or spoken words) that could be just passively downloaded; we rather attribute meanings to texts and to speech in the light of what we already know and believe. People's actual religious thought and behaviour are thus, for the most part, not based on making applied deductions from theological systems.

Underlying this is a view of culture as a 'precipitation of cognition and communication in a human population' (Sperber 1996, p. 97). The cultural and the psychological are not so much two different levels as measures of the spread of concepts and beliefs (Sperber 2006). Psychology is relevant when we want to explain how widespread concepts and beliefs are mentally represented and processed. This processing is to a large extent non-conscious: we do not have access to our own intuitive cognitive processes. Cognitive scientists of religion speak of 'intuitions' when they refer to such non-conscious beliefs and judgments (Atran 2002; Barrett, 2004; Boyer 1994, 2001; Pyysiäinen 2004a). In the so-called dual-process theories in social and neuropsychology as well as cognitive science, intuitive processing is distinguished from rational or systematic processing; explanations as to how these two types of processes or systems operate and are neurally processed vary, however (see Evans 2008; Evans and Frankish 2009; Pyysiäinen 2004a, 2009, pp. 189–92; Tremlin 2006, pp. 172–82).

Emphasis on intuitions means that research targets religious concepts and beliefs as they are actually used in making inferences in everyday life. There is experimental evidence that when persons make fast 'on-line' inferences from religious concepts they tend to replace their explicitly held, conscious beliefs by tacit intuitions: God, for example, can answer a prayer only after having finished answering another prayer, although at the explicit level God is reported as knowing everything simultaneously (Barrett 1998; Barrett and Keil 1996). Such distortion of theology happens because in fast and spontaneous reasoning intuitions intrude on explicit reasoning, without being voluntarily brought to the chain of inferences.

Theologically correct beliefs are difficult to use and spread because often it is not possible to make relevant spontaneous inferences from them (Boyer 2001, p. 321). Thus, theological elaboration of folk-religious concepts easily leads to a loss of relevance, which may then lead to the emergence of various kinds of revivalist movements emphasizing personal faith (see Poloma 1997; Pyysiäinen 2004c). Such fluctuation between theologically correct and more intuitive beliefs is reflected in, for example, interpretations of ritual efficacy; people have strong intuitions about supernatural agency as the key element in religious rituals, irrespective of theologically correct beliefs (Barrett and Lawson 2001). Rituals can be understood either as 'magically' bringing about changes in the natural world or as having effects on the supernatural world by, for example, changing people's relationship with God (Pyysiäinen 2004b, pp. 135–46; Sørensen 2007, pp. 5–6, 174–5, 188).

If either one of these roles of rituals is overemphasized at the cost of the other, the result is an imbalanced system where either relevance or unifying doctrinal power is lost (McCauley and Lawson 2002, pp. 184–92). While the possibility of making intuitive inferences makes religion persist, theological elaboration stabilizes explicit interpretative processes and thus creates coherence and unity in the group in question. There is evidence that large and complex societies are more likely to subscribe to potent deities directly concerned with morality; large, complex societies thus seem to have coevolved with (mono)theistic doctrines (Atran 2010; Atran and Henrich 2010). Yet theology cannot operate without individuals with enough motivation that comes from the practical applicability of religious concepts.

The intuitive inferences are based on intuitive ontology where personal agents form an especially important category. Counterintuitive agents, such as gods, spirits, elves, werewolves, and so forth are unusual members of the category of personal agents as some of their properties violate the intuitive expectations concerning the category of personal agents. Normally, agents, for example, have a body, but God is a bodiless spirit. Boyer (1994, 2001, 2003; see Knight et al. 2004) refers to such violations as 'counterintuitive'. Counterintuitiveness is based on our capacity to imagine entities that do not quite correspond to any intuitive category. Minimally counterintuitive representations involve only one violation of intuitive expectations, whereas everything else remains intuitive (Atran 2002; Barrett 2004; Boyer 1994, 2001; Pyysiäinen 2009).

There is cross-cultural experimental evidence to support the hypothesis that minimally counterintuitive representations are more memorable than intuitive or massively counterintuitive representations; therefore, they are more likely to become widespread within and across cultures, and narratives with some counterintuitive elements are easily remembered (Atran and Norenzayan 2004; Barrett and Nyhof 2001; Boyer and Ramble 2001; Gonce et al. 2006; Norenzayan and Atran 2004; Pyysiäinen et al. 2003; Tweney et al. 2006; Upal 2005, 2011; Upal et al. 2007). It is, however, no easy task to determine whether a representation involves one or more violations, although a way of measuring counterintuitiveness has recently been developed by Barrett (2008a).

AGENCY

Agency is one of the key concepts in the cognitive science of religion (Lawson 2001). There is experimental evidence to the effect that people associate counterintuitive agency with 'religion' (Pyysiäinen et al. 2003) and that an appropriate agent is the single most important factor in the efficacy of religious rituals (Barrett and Lawson 2001). Also supernatural agents are agents in the sense of organisms whose behaviour can be predicted by attributing to them conscious beliefs and desires (Dennett 1993, pp. 15–17), or whose behaviour is actually caused by their own mental states (Bechtel 2008, p. xii). Agency consists of the components of ANIMACY (liveliness, self-propelledness), and MENTALITY (beliefs and desires) (Pyysiäinen 2009, pp. 13–41). Animacy refers to liveliness expressed as self-propelledness. Only moving entities have a mind and consciousness, which thus seem to have their root in motor action (see Gallese and Metzinger 2003).

Detecting agency from motion cues, by and large irrespective of the specific form of the moving entities, has been supposed to be the precursor of our ability to attribute mental states to others (mentalizing or mind-reading). There is, however, evidence that brain areas such as the superior temporal sulcus and the temporo-parietal juncture are most responsive when subjects view real people and not, for example, animated cartoon characters. There seem to be neural circuits that respond specifically to patterns of intentional *biological* motion (Mar et al. 2007). But it may also be that the temporo-parietal junction is sensitive not only to biological motion but also to any stimuli which signal intentional activity (Gallagher et al. 2000). It has recently been found that even virtual agency represented as digital images on screen can trigger the mirror neuron system and spontaneous interpretation of agent causality (Howard-Jones et al. 2010). In another recent experiment, animal and non-living stimuli were processed in different parts of the brain in both blind and sighted persons; consequently, this categorization must be part of the hard-wired organization of brain and independent of visual experience (Mahon et al. 2009). It has also been shown that an animated figure that closely resembles but is not quite human is disturbing and even felt to be physically revolting (Hari and Kujala 2009, p. 460).

Yet it is obvious that quite minimal cues can lead us to postulate agency to account for perceived events (Heider and Simmel 1944). Guthrie (1993) argues that the fact that we see 'faces in clouds' and tend to animate the world around us is the immediate cause of religion; religion is a systematic form of anthropomorphism. The existence of a 'hypersensitive agent detection device' (HADD) in the human mind has subsequently become one of the core hypotheses in the standard model of the cognitive science of religion (Barrett 2000, 2004; Boyer 2005b), although there is little empirical research on this issue (but see Barrett and Johnson 2003). It has, however, been suggested that individual differences in the activation level of HADD might explain why some persons are prone to atheism and some to belief in supernatural agency (Saler and Ziegler 2006).

Boyer and Barrett (2005) argue that the following cues lead to the inference of agency:

(1) Animate motion that has as its input such things as non-linear changes in direction, sudden acceleration without collision, and change of physical shape that accompanies motion (e.g. caterpillar-like crawling);

(2) An object reacting at a distance;

(3) Trajectories that only make sense on the condition that the moving entity is trying to reach or avoid something, which leads to goal-ascription;

(4) The end-result of an action being connected to perceived movement through the link of the intention of the moving organism (intention-ascription);
(5) The experience of joint attention for which we develop a capacity between nine and twelve months of age.

Yet it may be that the first three are only cues about animacy while the last two are cues for postulating mentality (Pyysiäinen 2009, p. 14). In any case, beliefs about supernatural agents cannot be explained merely as due to the false positives that HADD produces; such beliefs are widespread and persist even in the absence of perceptual cues (Barrett 2004, p. 38; Pyysiäinen 2009, p. 22). Normally, we discard the false positives produced by HADD as soon as we recognize them as such; religious beliefs, however, persist and do not necessarily presuppose any objective evidence to support them. They persist because they feel compelling and because it is possible to make relevant inferences from them. In other words, god beliefs are not important because people believe in them; people rather believe in them because they are important (Boyer 2001, pp. 298–304; Pyysiäinen 2003). Being widespread and being an object of belief are two different things, though (Pyysiäinen 2003). Mere cognitive structure is not enough to explain belief; people do not believe in alien gods even when the cognitive structure of their mental representation is quite similar to that of the representation of one's own gods (the 'Zeus problem'; Gervais and Henrich 2010). Phenomena such as imitation of the majority or of prestigious individuals seem to be important causal factors in the adoption of belief (see Henrich and Henrich 2007; Richerson and Boyd 2005).

In asking how hypersensitive agent detection produces supernatural agent concepts, Atran and Henrich (2010) argue that four things are needed: dynamic interaction of the by-products of adaptive cognitive mechanisms, adaptive learning heuristics like emulation of prestigious individuals or the majority, credibility-enhancing ritual displays, and cultural group selection. Cultural evolution has favoured the emergence of beliefs in powerful moralizing deities concerned with the prosocial behaviour of individuals beyond kin- and reciprocity-based networks. Modern large-scale religions thus have evolved to create a potent link between the supernatural and the prosocial (see Norenzayan and Shariff 2008; Sanderson and Roberts 2008).

THE PROGRAMME AND ITS PROBLEMS

The leading idea of the cognitive science of religion is the already mentioned 'epidemiology of beliefs' (Sperber 1985, 1996): religious concepts and beliefs are widespread because all healthy humans have cognitive mechanisms that make it easy for them to remember and process in mind concepts and beliefs generally deemed 'religious'. Such concepts and beliefs are 'contagious', as it were. The epidemiological account of religion is a general-level theory that seeks to explain the spread of religious beliefs, not their historical or psycho-dynamic origins. It still needs to be supported by more experimental evidence, however (see Claidière and Sperber 2007). Individual studies on the historical spread of particular religious concepts and beliefs are also by and large still missing (cf. Pyysiäinen 2009). As Barrett puts it, cognitive theories have not been applied to *particular* problems, scholars rather studying 'why religious rituals appear the way they do *generally*, why people believe in gods generally', and so forth. This is also often accompanied by attempts at solving only theoretical problems by conceptual analysis alone (Barrett 2008b, p. 298).

As an attempt to interpret historical data or to theorize about religion at a general level, the cognitive science of religion is one perspective in the study of religion or comparative religion. A rich and empirically tested theoretical apparatus is used to theorize about religious belief and behaviour. Notwithstanding the danger of only creating 'just so stories', the cognitive science of religion is a legitimate and important area of research. However, if research on religion is reduced to the experimental study of the neurocognitive mechanisms underlying religion, the whole study of religion is reduced to general cognitive and neuroscience (see Pyysiäinen, in press b). This does not mean that good work on religion could not be done along these lines; it only means that the existence of the study of religion as an independent discipline becomes suspect.

But it might also be possible to understand the cognitive science of religion as a forum where experimental psychologists, evolutionary theorists, anthropologists, and religion scholars can meet, exchange ideas, and create multidisciplinary projects (see Barrett 2008b, p. 297; Gibson and Barrett 2008). Such cross-fertilization of ideas is extremely important because the traditional academic boundaries between disciplines often make truly interdisciplinary work difficult and because religion is such a heterogeneous category and can be approached at differing levels (see Boyer 2005a; Pyysiäinen, in press a). The cognitive science of religion thus can serve as an alternative for obsolete essentialist views of religion as a *sui generis* category. Although there is no universal essence of religion, religious beliefs, and behaviours yet form an important element in the evolution of adaptive prosocial behaviour that makes intragroup cooperation possible but at the same time often creates conflicts between groups with differing beliefs (see Atran 2010).

CONCLUSION

The cognitive science of religion has by and large focused on how human cognitive architecture has made religious concepts possible and easy to spread within and across cultures. The question of why and how these concepts and beliefs have become objects of serious belief cannot be answered by mere cognitive considerations alone. We must also take into consideration sociocultural processes such as imitation and group dynamics as well as ritual behaviour. Commitment to religion depends on costly and hard to fake signalling of sincere commitment to shared values and norms (e.g. Atran 2010; Bulbulia and Schjoedt 2010). The human cognitive architecture and capacity for emotional reactions make this possible, but these are intertwined with such group dynamics that make individuals prone to follow norms and values of the majority or of prestigious individuals. Cooperation within large groups of 'fictive kin' is made possible by beliefs about all-knowing gods that help tie together genetically unrelated individuals into a cooperative community. (Mono)theistic religions are a means of creating and sustaining cooperation among genetically non-related individuals in complex large-scale societies.

The cognitive science of religion has drawn the study of religion from its relative isolation to the interdisciplinary field of sociocognitive and behavioural sciences. To the extent that religion is not a category *sui generis*, its study can shed light also on features of human behaviour in general. There is no essence of religion, 'religion' rather being a scholarly construct we use to lump together a variety of phenomena between which there is a family resemblance; without that there would not be any necessary links between the various aspects of 'religion' (Boyer 2010; Saler 2000). Religion can be studied at various levels of analysis from neurochemistry to mind and culture. The questions we want to ask determine the level at which we seek for answers. Thus there is no predetermined 'right' level of analysis. By the same token, concepts such as 'transcendence' and the 'holy' are significant only to the extent that they can be operationalized to serve the purposes of empirical research. This does not require an ontological stance with regard to the existence or non-existence of gods; whether such entities exist or not, we study only how humans take their supposed existence into account in their behaviour.

Such research has often been accused of reductionism, but from the point of view of philosophy of science the whole notion of reduction has become obsolete: all theories and explanations are reductionist from some point of view. It is the *explanandum* in question that determines at which level of analysis we seek for answers: neurochemistry at the molecular level, neural networks, cognitive architecture, or the sociocultural (Craver 2007; McCauley 2007; Pyysiäinen 2012). The cognitive science of religion mainly focuses on one level: cognitive mechanisms. These, however, also have a bearing on cultural-level phenomena, the cultural and the cognitive being measures of the distribution of mental representations rather than two distinct spheres. An optimal strategy in research thus is multi-level mechanistic explanation that specifies the various sorts of mechanisms (neural, cognitive, etc.) that produce and support religious phenomena at varying ontological levels (see Bechtel 2008; Craver 2007; McCauley 2007; Pyysiäinen 2012, in press a).

REFERENCES

Atkinson, Q. D. and P. Bourrat. 2011. Beliefs about God, the afterlife and morality support the role of supernatural policing in human cooperation. *Evolution and Human Behavior 32*: 41–9.

Atran, S. 1987. Ordinary constraints on the semantics of living kinds. *Mind and Language 2*: 271–95.

——. 2002. *In Gods We Trust: The Evolutionary Landscape of Religion*. Oxford: Oxford University Press.

——. 2010. *Talking to the Enemy: Faith, Brotherhood and the (Un)making of Terrorists*. New York: Harper Collins.

Atran, S. and A. Norenzayan. 2004. Religion's evolutionary landscape: Counterintuition, commitment, compassion, communion. *Behavioral and Brain Sciences 27*(6): 713–30 [with commentaries].

Atran, S. and D. Medin. 2008. *The Native Mind and the Cultural Construction of Nature*. Cambridge, MA: MIT Press.

Atran, S. and J. Henrich. 2010. The evolution of religion: How cognitive by-products, adaptive learning heuristics, ritual displays, and group competition generate deep commitments to prosocial religions. *Biological Theory 5*(1): 18–30.

Axelrod, R. 1990/1984. *The Evolution of Co-operation*. Harmondsworth: Penguin.

Barrett, J. L. 1998. Cognitive constraints on Hindu concepts of the divine. *Journal for the Scientific Study of Religion 37*(4): 608–19.

——. 1999. Theological correctness: Cognitive constraint and the study of religion. *Method & Theory in the Study of Religion 11*: 325–39.

——. 2000. Exploring the natural foundations of religion. *Trends in Cognitive Sciences 4*: 29–34.

——. 2004. *Why Would Anyone Believe in God*? Walnut Creek, CA: AltaMira Press.

——. 2007. Cognitive science of religion: What is it and why is it? *Religion Compass 2*. doi:10.1111/j.1749-8171.2007.00042.x.

——. 2008a. Coding and quantifying counterintuitiveness: Theoretical and methodological reflections. *Method & Theory in the Study of Religion 20*(4): 308–38.

——. 2008b. Keeping 'science' in the cognitive science of religion. In *The Evolution of Religion: Studies, Theories, & Critiques*. Eds. J. Bulbulia, R. Sosis, E. Harris, R. Genet, C. Genet, and K. Wyman. Santa Margarita, CA: Collins Family Foundation, 295–301.

——. 2011. *Cognitive Science, Religion, and Theology: From Human Minds to Divine Minds. (Templeton Science and Religion Series.)* Pennsylvania: Templeton Press.

Barrett, J. L. and F. Keil. 1996. Conceptualizing a nonnatural entity: Anthropomorphism in God concepts. *Cognitive Psychology 31*: 219–47.

Barrett, J. L. and M. Nyhof. 2001. Spreading non-natural concepts: The role of intuitive conceptual structures in memory and transmission of cultural materials. *Journal of Cognition and Culture 1*: 69–100.

Barrett, J. L., R. A. Richert, and A. Driesenga. 2001. God's beliefs versus mothers: The development of nonhuman agent concepts. *Child Development 72*(1): 50–65.

Barrett, J. L. and E. T. Lawson. 2001. Ritual intuitions: Cognitive contributions to judgments of ritual efficacy. *Journal of Cognition and Culture 1*(2): 183–201.

Barrett, J. L. and A. H. Johnson. 2003. The role of control in attributing intentional agency to inanimate objects. *Journal of Cognition and Culture* 3(3): 208–17.

Bateson, M., D. Nettle, and G. Roberts. 2006. Cues of being watched enhance cooperation in a real-world setting. *Biology Letters* 2: 412–14. doi:10.1098/rsbl.2006.0509.

Bechtel, W. 2008. *Mental Mechanisms: Philosophical Perspectives on Cognitive Neuroscience*. New York: Routledge.

Beck, R. 2006. *The Religion of the Mithras Cult in the Roman Empire: Mysteries of the Unconquered Sun*. Oxford: Oxford University Press.

Bering, J. M. 2006. The folk psychology of souls. *Behavioral and Brain Sciences* 29(5): 453–62.

——. 2011. *The Belief Instinct*. New York: Norton.

Bering, J. M. and D. F. Bjorklund. 2004. The natural emergence of 'afterlife' reasoning as a developmental regularity. *Developmental Psychology* 40(2): 217–33.

Bering, J. M. and D. D. P. Johnson. 2005. 'Oh Lord...you perceive my thoughts from afar': Recursivenes and the evolution of supernatural agency. *Journal of Cognition and Culture* 5(1–2): 118–42.

Bering, J. M. and B. D. Parker. 2006. Children's attribution of intentions to an invisible agent. *Developmental Psychology* 42(2): 253–62.

Boyer, P. 1990. *Tradition as Truth and Communication*. Cambridge: Cambridge University Press.

——. 1994. *The Naturalness of Religious Ideas: A Cognitive Theory of Religion*. Berkeley: University of California Press.

——. 2001. *Religion Explained: The Evolutionary Origins of Religious Thought*. New York: Basic Books.

——. 2003. Religious thought and behaviour as byproducts of brain function. *Trends in Cognitive Sciences* 7(3): 119–24.

——. 2005a. Ten problems for integrated behavioural science: How to make the social sciences relevant. *Global Fellows Seminars, University of California, Los Angeles; paper #7*. Online; accessed in February 2005. Available online at <http://repositories.cdlib.org/globalfellows/2005/7>

——. 2005b. A reductionistic model of distinct modes of religious transmission. In *Mind and Religion: Psychological and Cognitive Foundations of Religiosity*. Eds. H. Whitehouse and R. N. McCauley, 3–29. Walnut Creek, CA: AltaMira Press.

——. 2010. *The Fracture of an Illusion: Science and the Dissolution of Religion*. Göttingen: Vandenhoeck and Ruprecht.

Boyer, P. and C. Ramble. 2001. Cognitive templates for religious concepts: Cross-cultural evidence for recall of counterintuitive representations. *Cognitive Science* 25: 535–64.

Boyer, P. and H. C. Barrett. 2005. Domain-specificity and intuitive ontology. In *The Handbook of Evolutionary Psychology*. Ed. D. Buss. Hoboken, NJ: John Wiley, 96–118.

Boyer, P. and P. Liénard. 2006. Why ritualized behavior? Precaution systems and action-parsing in developmental, pathological and cultural rituals. *Behavioral and Brain Sciences* 29(6): 595–650 [with commentaries].

Boyer, P. and B. Bergstrom. 2008. Evolutionary perspectives on religion. *Annual Review of Anthropology* 37: 111–310.

Bulbulia, J. and U. Schjoedt. 2010. Religious culture and cooperative prediction under risk: Perspectives from social neuroscience. In *Religion, Economy, and Cooperation*. Ed. I. Pyysiäinen. (*Religion and Reason*, 49.) Berlin: Mouton de Gruyter, 35–59.

Claidière, N. and D. Sperber. 2007. The role of attraction in cultural evolution: Reply to J. Henrich and R. Boyd, 'On modelling cognition and culture'. *Journal of Cognition and Culture* 7(1–2): 89–111.

——. 2010. Imitation explains the propagation, not the stability of animal culture. *Proceedings of the Royal Society B: Biological Sciences 277*(1681): 651–9.

Craver, C. 2007. *Explaining the Brain: Mechanisms and the Mosaic Unity of Neuroscience*. New York: Oxford University Press.

Czachesz, I. 2007. The transmission of early Christian thought: Toward a cognitive psychological model. *Studies in Religion 36*: 65–84.

Day, M. 2009. Exotic experience and ordinary life: On Andrew Newberg, Eugene d'Aquili and Vince Rause, *Why God Won't Go Away*. In *Contemporary Theories of Religion: A Critical Companion*. Ed. M. Stausberg. London: Routledge, 115–28.

Dennett, D. C. 1993/1987. *The Intentional Stance*. Cambridge, MA: MIT Press.

——. 1991. *Consciousness Explained*. Harmondsworth: Penguin.

Depew, D. J. 2003. Baldwin and his many effects. In *Evolution and Learning: The Baldwin Effect Reconsidered*. Eds. B. H. Weber and D. J. Depew. Cambridge, MA: MIT Press, 3–30.

Evans, J. St. B.T. 2008. Dual-processing accounts of reasoning, judgment, and social cognition. *Annual Review of Psychology 59*: 255–78.

Evans, J. St. B.T. and K. Frankish. Eds. 2009. *In Two Minds: Dual Processes and Beyond*. Oxford: Oxford University Press.

Gallagher, H. L., F. Happé, N. Brunswick, P. C. Fletcher, U. Frith, and C. D. Frith. 2000. Reading the mind in cartoons and stories: An fMRI study of 'theory of mind' in verbal and nonverbal tasks. *Neuropsychologia 38*: 11–21.

Gallese, V. and T. Metzinger. 2003. Motor ontology: The representational reality of goals, actions and selves. *Philosophical Psychology 16*(3): 365–88.

Gervais, W. and J. Henrich. 2010. The Zeus problem: Why representational content biases cannot explain faith in gods. *Journal of Cognition and Culture 10*: 383–9.

Gibson, N. and J. Barrett. 2008. On psychology and evolution of religion: Five types of contribution needed from psychologists. In *The Evolution of Religion: Studies, Theories, & Critiques*. Eds. J. Bulbulia, R. Sosis, E. Harris, R. Genet, C. Genet, and K. Wyman. Santa Margarita, CA: Collins Family Foundation, 333–8.

Godfrey-Smith, P., D. Dennett, and T. W. Deacon. 2003. Postscript on the Baldwin effect and niche construction. In *Evolution and Learning: The Baldwin Effect Reconsidered*. Eds. B. H. Weber and D. J. Depew. Cambridge, MA: MIT Press, 107–12.

Gonce, L. O., A. Upal, D. J. Slone, and R. D. Tweney. 2006. Role of context in recall of counterintuitive concepts. *Journal of Cognition and Culture* 6(3–4): 521–47.

Gopnik, A., A. N. Meltzoff, and P. K. Kuhl. 1999. *The Scientist in the Crib*. New York: Harper Collins.

Guthrie, S. 1993. *Faces in the Clouds*. New York: Oxford University Press.

Hamilton, W.D. 1964. The genetic evolution of social behaviour I, II. *Journal of Theoretical Biology 7*: 1–16, 17–52.

Hari, R. and M. Kujala. 2009. Brain basis of human social interaction: From concepts to brain imaging. *Physiological Reviews 89*: 453–479.

Heider, F. and M.-A. Simmel. 1944. An experimental study of apparent behavior. *American Journal of Psychology 57*: 243–9.

Henrich, N. and J. Henrich. 2007. *Why Humans Cooperate: A Cultural and Evolutionary Explanation*. Oxford: Oxford University Press.

Hirschfeld, L. A. and S. A. Gelman. Eds. 1994. *Mapping the Mind*. Cambridge: Cambridge University Press.

Howard-Jones, P. A., R. Bogacz, J. H. Yoo, U. Leonards, and S. Demetriou. 2010. The neural mechanisms of learning from competitors. *Neuroimage 53*(2): 790–9.

Johnson, D. D. P. 2011. Why God is the best punisher. *Religion, Brain & Behavior 1*(1): 77–84.

Keil, F.C. 1979. *Semantic and Conceptual Development*. Cambridge, MA: Harvard University Press.

——. 1996/1989. *Concepts, Kinds, and Cognitive Development*. Cambridge, MA: MIT Press.

Knight, N., P. Sousa, J. L. Barrett, and S. Atran. 2004. Children's attributions of beliefs to humans and God: Cross-cultural evidence. *Cognitive Science 28*: 117–26.

Lawson, E. T. 2001. Psychological perspectives on agency. In *Religion in Mind: Cognitive Perspectives on Religious Belief, Ritual and Experience*. Ed. J. Andresen. Cambridge: Cambridge University Press, 141–72.

Lawson, E. T. and R. N. McCauley. 1990. *Rethinking Religion: Connecting Cognition and Culture*. Cambridge: Cambridge University Press.

Liénard, P. and P. Boyer. 2006. Whence collective rituals? A cultural selection model of ritualized behavior. *American Anthropologist 108*(4): 814–27.

Luomanen, P., I. Pyysiäinen, and R. Uro. Eds. 2007. *Explaining Christian Origins and Early Judaism: Contributions from Cognitive and Social Science* (*Biblical Interpretation Series*, 89). Leiden: Brill.

McCauley, R. N. 2007. Reduction: Models of cross-scientific relations and their implications for the psychology-neuroscience interface. In *Philosophy of Psychology and Cognitive Science*. Ed. P. Thagard. (*Handbook of the Philosophy of Science Series*.) Amsterdam: North Holland/Elsevier, 105–58.

——. 2011. *Why Religion is Natural and Science is Not*. New York: Oxford University Press.

McCauley, R. N. and E. T. Lawson. 2002. *Bringing Ritual to Mind: Psychological Foundations of Cultural Forms*. Cambridge: Cambridge University Press.

Mahon, B. Z., S. Anzellotti, J. Schwarzback, and M. Zampini. 2009. Category-specific organization in the human brain does not require visual experience. *Nature 63* (August 20): 397–405.

Mar, R. A., W. M. Kelley, T. F. Heatherton, and C. N. Macrae. 2007. Detecting agency from the biological motion of veridical *vs* animated agents. *Social Cognitive and Affective Neuroscience* (*SCAN*) *2*: 199–205. Doi:10.1093/scan/nsm011.

Martin, L. H. 2003. Cognition, society and religion: A new approach to the study of culture. *Culture and Religion 4*(2): 207–31.

——. 2004. Towards a cognitive history of religions. In *Unterwegs. Neue Pfade in der Religionswissenschaft / New Paths in the Study of Religions: Festschrift for Michael Pye*. Eds. C. Kleine, M. Schrimpf, and K. Triplett. München: Biblion, 75–82.

Norenzayan, A. and S. Atran. 2004. Cognitive and emotional processes in the cultural transmission of natural and nonnatural beliefs. In *The Psychological Foundations of Culture*. Eds. M. Schaller and C. Crandall. Mahwah, NJ: Lawrence Erlbaum Associates, 149–69.

Norenzayan, A. and A. F. Shariff. 2008 The origin and evolution of religious prosociality. *Science 322*: 58–62.

Poloma, M. 1997. The 'Toronto Blessing:' Charisma, institutionalization, and revival. *Journal for the Scientific Study of Religion 36*: 257–71.

Pyysiäinen, I. 2001. *How Religion Works: Towards a New Cognitive Science of Religion*. Leiden: Brill.

——. 2003. True fiction: Philosophy and psychology of religious belief. *Philosophical Psychology 16*(1): 109–25.

——. 2004a. Intuitive and explicit in religious thought. *Journal of Cognition and Culture 4*(1): 123–50.

——. 2004b. *Magic, Miracles, and Religion: A Scientist's Perspective*. Walnut Creek, CA: AltaMira Press.

——. 2004c. Corrupt doctrine and doctrinal revival: On the nature and limits of the modes theory. In *Theorizing Religions Past: Archaeology, History, and Cognition*. Eds. H. Whitehouse and L. H. Martin, 173–94. Walnut Creek, CA: AltaMira Press.

——. 2006. Amazing grace: Religion and the evolution of the human mind. In *Where God and Science Meet: How Brain and Evolutionary Studies Alter Our Understanding of Religion, I–III. Volume I, Evolution, Genes, and the Religious Brain*. Ed. P. McNamara. Westport, CT: Praeger, 209–25.

——. 2008. Introduction: Religion, cognition, and culture. *Religion 38*(2): 101–8.

——. 2009. *Supernatural agents: Why We Believe in Souls, Gods, and Buddhas*. New York: Oxford University Press.

——. 2012. Religion: From mind to society and back. In *Grounding Social Sciences in Cognitive Sciences*. Ed. R. Sun. Cambridge, MA: MIT Press.

——. (In press a). Mechanical explanation of ritualized behavior. In *Religious Ritual, Cognition and Culture*. Eds. A. Geertz and J. Sørensen. London: Equinox.

——. (In press b). Religion and the brain—Cognitive science as a basis for theories of religion. In *Theory / Religion / Critique: Classic and Contemporary Approaches*. Ed. R. King. New York: Columbia University Press.

Pyysiäinen, I., M. Lindeman, and T. Honkela. 2003. Counterintuitiveness as the hallmark of religiosity. *Religion 33*(4): 341–55.

Pyysiäinen, I. and M. Hauser. 2010. The origins of religion: Evolved adaptation or by-product? *Trends in Cognitive Sciences 14*(3): 104–9.

Richerson, P. and R. Boyd. 2005. *Not by Genes Alone: How Culture Transformed Human Evolution*. Chicago: University of Chicago Press.

Saler, B. 2000/1993. *Conceptualizing Religion: Immanent Anthropologists, Transcendent Natives, and Unbound Categories*. [With a new preface.] New York: Berghahn Books.

Saler, B. and C.A. Ziegler. 2006. Atheism and the apotheosis of agency. *Temenos 42*(2): 7–41.

Sanderson, S. K. and W. W. Roberts. 2008 The evolutionary forms of religious life: A cross-cultural, quantitative analysis. *American Anthropologist 110*: 454–66.

Schloss, J. P. and M. Murray. 2011. Evolutionary accounts of beliefs in supernatural punishment: A critical review. *Religion, Brain & Behavior 1*(1): 46–99.

Shariff, A. 2011. Big gods were made for big groups. *Religion, Brain & Behavior 1*(1): 89–93.

Sørensen, J. 2007. *A Cognitive Theory of Magic*. Lanham, MD: AltaMira.

Sperber, D. 1995/1975. *Rethinking Symbolism*. Cambridge: Cambridge University Press.

——. 1985. Anthropology and psychology: Towards an epidemiology of representations. *Man* (N.S.) *20*: 73–89.

——. 1996. *Explaining culture: A naturalistic approach*. Oxford: Blackwell.

——. 2006. Conceptual tools for a naturalistic approach to cultural evolution. In *Evolution of Culture: A Fyssen Foundation Symposium*. Eds. S. C. Levinson and P. Jaisson. Cambridge, MA: MIT Press, 147–65.

Sperber, D. and D. Wilson. 1988/1986. *Relevance: Communication and Cognition*. Cambridge, MA: Harvard University Press.

Sperber, D., D. Premack, and A. J. Premack. Eds. 1995. *Causal Cognition*. Oxford: Clarendon Press.

Thagard, P. 2005. The emotional coherence of religion. *Journal of Cognition and Culture*, *5*(1–2): 58–74.

Tremlin, T. 2006. *Minds and Gods: The Cognitive Foundations of Religion*. New York: Oxford University Press.

Trivers, R. 2002. *Natural Selection and Social Theory: Selected Papers of Robert Trivers*. Oxford: Oxford University Press.

Tweney, R. M., A. Upal, L. O. Gonce, D. J. Slone, and K. Edwards. 2006. The creative structuring of counterintuitive worlds. *Journal of Cognition and Culture* *6*(3–4): 483–98.

Upal, A. 2005. Role of context in memorability of intuitive and counterintuitive concepts. In *Proceedings of 27th Annual Meeting of the Cognitive Science Society*. Mahwah; NJ: Erlbaum, 2224–9.

——. 2011. Memory, mystery and coherence: Does the presence of 2–3 counterintuitive concepts predict cultural success of a narrative? *Journal of Cognition and Culture* *11*(1–2): 23–48.

Upal, A., L. O. Gonce, D. J. Slone, and R. D. Tweney. 2007. Contextualizing counterintuitiveness: How context affects comprehension and memorability of counterintuitive concepts. *Cognitive Science 31*: 1–25.

Visala. A. 2011. *Naturalism, Theism and the Cognitive Study of Religion: Religion Explained? (Ashgate Science and Religion Series)*. Farnham: Ashgate.

Whitehouse, H. 1995. *Inside the Cult: Religious Innovation and Transmission in Papua New Guinea*. Oxford: Clarendon Press.

——. 2004. *Modes of Religiosity: A Cognitive Theory of Religious Transmission*. Walnut Creek, CA: AltaMira Press.

Wilson, D. S. 2002. *Darwin's Cathedral: Evolution, Religion and the Nature of Society*. Chicago: University of Chicago Press.

3

Biologically Evolutionary Explanations of Religious Belief

MICHAEL RUSE

Does religion have a natural origin and is that natural origin explicable in terms of evolutionary biology? And if this is so, does this have any implications for the truth of religious beliefs? These are the questions of this essay. They are questions that have been asked many times before. Most recently, they have commanded a lot of attention from cognitive scientists, evolutionary psychologists and others in various fields who suppose that recent research in the cognitive science of religion might help us understand the evolutionary history of religion. This enterprise, which I take to be exemplified in the work of Pascal Boyer and others, is clearly novel in some respects, but it can also be seen as the latest in a long line of evolutionary explanations of one kind or another. Each of these, it might be argued, poses a potential threat to the truth of religion, and each has failed adequately to resolve a number of age-old debates (for example, debates about the adaptiveness or non-adaptiveness of religion, and whether religion is best explained in terms of biological or cultural evolution).

My aim, here, is to give an overview of some of the key debates that have characterized the long history of the evolutionary study of religion. Throughout, I will be particularly concerned with questions about the extent to which evolutionary biology can help answer questions about the natural origins of religion, and with how Darwinian thought has been variously employed, developed, and/or displaced by those offering different kinds of evolutionary explanation of religion. By so doing, I hope to set the evolutionary cognitive science of religion in its historical context, though I will not dwell on the finer details of this field (Ilkka Pyysiäinen surveys the key debates admirably in the previous chapter). In describing ECSR's historical background, I hope to show the enduring nature of the evolutionary debates in which it is immersed. It is no trivial matter, as all evolutionary theories of religion, including ECSR, must, at some point, address the crucial question of how much can really be explained by our biological heritage. I will not, however, attempt to give definitive answers to the difficult questions that pervade this complex history, but rather suggest that attitudes to these questions are likely to depend to a certain extent upon what kinds of evidence one takes seriously. In the final section of this essay I will examine the implications of claims about the natural evolutionary origins of religion for the truth status of religion.

DAVID HUME AND CHARLES DARWIN

Any account of the evolutionary origins of religious belief starts with David Hume. I do not think that Hume was an evolutionist; but, both in his own right and because of his influence on Charles Darwin, his essay, 'The Natural History of Religion', is the definitive text. In that essay, Hume argued that religious beliefs are entirely naturally caused and that they have a developmental history, one taking them from rather primitive polytheistic notions up to the more sophisticated religions of his own day. In a famous passage, he wrote:

> There is a universal tendency among mankind to conceive all beings like themselves, and to transfer to every object, those qualities, with which they are familiarly acquainted, and of which they are intimately conscious. We find human faces in the moon, armies in the clouds; and by a natural propensity, if not corrected by experience and reflection, ascribe malice or good-will to everything, that hurts or pleases us. ([1757] 1963, pp. 40–1)

Hume argued that this tendency led naturally to a belief that the world is filled with god or spirits of some kind. Then, he suggested that some gods start to win out over others, until finally we end up with a single god, monotheism. One should note that Hume was not particularly keen on the end result, and—undoubtedly showing his Calvinistic upbringing—he was particularly contemptuous of Roman Catholic beliefs such as transubstantiation, the belief that we are actually eating our god.

> Upon the whole, the greatest and most observable differences between a traditional, mythological religion, and a systematic, scholastic one are two: The former is often more reasonable, as consisting only of a multitude of stories, which, however groundless, imply no express absurdity and demonstrative contradiction; and sits also so easy and light on men's minds, that, though it may be as universally received, it happily makes no such deep impression on the affections and understanding. (p. 85)

There is ongoing discussion among Darwin scholars as to the links, direct or indirect, between the ideas of David Hume and Charles Darwin. However, on the matter of religion, there seems little doubt that there was direct and important influence. We know that Darwin read Hume's essay around the time that he became an evolutionist, that is to say in the late 1830s (Ruse 2008). When Darwin turned to explicit discussion of religion, in his *Descent of Man* (published in 1871), the treatment was almost word for word that of Hume. In particular, Darwin was eager to suggest that beliefs in God are not in any sense hardwired, so that we should at once think that they have adaptive virtues. He pointed out that there are many instances of human races where the idea of God is completely unknown.

> There is no evidence that man was aboriginally endowed with the ennobling belief in the existence of an Omnipotent God. On the contrary there is ample evidence, derived not from hasty travellers, but from men who have long resided with savages, that numerous races have existed, and still exist, who have no idea of one or more gods, and who have no words in their languages to express such an idea. (1871, 1, p. 65)

How then did Darwin think that religion came about? Very much along the lines that Hume had suggested, that is almost accidentally as the result of a tendency that is a by-product of our (directly selection-caused) human nature.

> The tendency in savages to imagine that natural objects and agencies are animated by spiritual or living essences, is perhaps illustrated by a little fact which I once noticed: my dog, a full-grown and very sensible animal, was lying on the lawn during a hot and still day; but at a little distance a slight breeze occasionally moved an open parasol, which would have been wholly disregarded by the dog, had any one stood near it. As it was, every time that the parasol slightly moved, the dog growled fiercely and barked. He must, I think, have reasoned to himself in a rapid and unconscious manner, that movement without any apparent cause indicated the presence of some strange living agent, and that no stranger had a right to be on his territory. (1, p. 67)

Like Hume, Darwin then went on to suggest that we move from primitive religion up to sophisticated religions, and in particular from belief in many gods (polytheism) to just one god (monotheism). Like Hume, Darwin suggested that even sophisticated religions—perhaps most particularly sophisticated religions—have little to commend them. What Darwin was keen to emphasize was that a belief in the natural origins of religion should not thereby commit one to believe that God necessarily exists. Summing up his thoughts on religion Darwin wrote:

> The belief in God has often been advanced as not only the greatest, but the most complete of all the distinctions between man and the lower animals. It is however impossible, as we have seen, to maintain that this belief is innate or instinctive in man. On the other hand a belief in all-pervading spiritual agencies seems to be universal; and apparently follows from a considerable advance in man's reason, and from a still greater advance in his faculties of imagination, curiosity and wonder. I am aware that the assumed instinctive belief in God has been used by

> many persons as an argument for His existence. But this is a rash argument, as we should thus be compelled to believe in the existence of many cruel and malignant spirits, only a little more powerful than man; for the belief in them is far more general than in a beneficent Deity. The idea of a universal and beneficent Creator does not seem to arise in the mind of man, until he has been elevated by long-continued culture. (2, pp. 394–5)

THE SOCIAL SCIENCES

The one hundred years from around the time of the *Descent of Man* to the early 1970s were something of a black hole for (biological) evolutionary explanations of human behaviours and beliefs and customs. There was little of value and what was produced tends today with good reason to be regarded with suspicion, not to say distaste. There were internal reasons for this and external reasons also. Internally, the most obvious reason is that evolutionary theory generally was in an immature state for much of the time (Ruse 1996). Darwin had no adequate theory of heredity—this came only at the beginning of the twentieth century with the rediscovery of Mendel's thinking on the subject—and it was the 1930s before evolutionists were able to take up the augmented theory of selection and move forward positively with a new paradigm (Ruse 2013). It was the 1960s and 1970s before they felt sufficiently confident to turn full attention to humankind, speaking now specifically at the social and cultural level (Ruse 1979). Even then there was the past to throw off, because as is well known Darwin's ideas had been appropriated by those who had social and political interests to promote. So-called 'Social Darwinism' became a byword for theses about the virtues of big industry, the justification of denying help to the poor, and even more dreadfully, the motives that led to the awful confrontations and social systems of the first half of the twentieth century. Externally came the rise of the social sciences and with this perhaps understandably an urge to deny the relevance of the biological sciences when it came to human nature. This was not just simply a fight for territory, but also a function of the greatly increased quality and quantity of empirical work, the diversity and complexity of which seemed to negate simple underpinnings in terms of natural selection and the growing theory of heredity. Intermingled with this no doubt even in the most secular of thinkers was an implicit belief that we humans are different. What may function as an adequate explanation for the behaviour of the warthog could never function as an adequate explanation for the behaviour of humankind.

Paradigmatic of the kind of approach taken and results obtained was the celebrated work of the Scottish anthropologist James George Frazer. His *Golden Bough: A Study in Magic and Religion* (first edition 1890, but much enlarged in later editions) is a full and comprehensive study of societies, their magical beliefs, and their religions. In a Comtian fashion, undoubtedly influenced by the evolutionary speculations of his day, seeing a succession in societies from magic, to religion, and then on to science, Frazer focused on phenomena that he found repeated from society to society, most notably fertility cults. Famously, or perhaps notoriously, he saw the key player as a king who must die, who is sacrificed, but who then reincarnates either himself or through a successor and who thereby ensures the good of the race and also in some versions the revival of nature in the spring after it has reached fulfilment in the previous autumn and then death in the winter. Naturally, he was not indifferent to the ways in which this tale, supposedly repeated in society after society, resonates with the Christian story, with the death of Jesus for our salvation and his subsequent revival at the moment of the Resurrection.

> If I am right, the motive force for slaying a man-god is a fear lest with the involvement of his body in sickness or old age his Sacred spirit should suffer a corresponding decay, which might imperil the general course of nature and with it the existence of his worshippers, who believed the cosmic energies to be mysteriously knit up with those of their human divinity. Hence, if there is any measure of truth in this theory, the practice of putting divine men and particularly divine kings to death, which seems to have been common at a particular stage in the evolution of society and religion, was a crude but pathetic attempt to disengage an immortal spirit from its mortal envelope, to arrest the forces of decomposition in nature by retrenching with ruthless hand the first ominous symptoms of decay. (1911, pp. v–vi)

It is naturalistic but entirely a matter of culture.

There were of course those who did want to meld biology with their cultural understanding of religion. One such was Sigmund Freud. It is now appreciated fully the extent to which he was committed to a Lamarckian theory of evolution, where the inheritance of acquired characters—the giraffe stretches its neck to get leaves and so the species evolves in that direction—plays the major role (Sulloway 1979). This commitment is fundamental to his theorizing about religion, well known for supposing that everything started with sons killing or castrating their father and then setting up incest taboos against having sex with their mother. One can generalize out from this, for instance by supposing that the killing was followed by an act of cannibalism where the father is eaten to gain his strength in some way and the incest taboo is seen as a foundation for the way in which religion constrains and channels all of our various desires to have our own way into acts that benefit society instead. Religion is thus totally illusory, and functions to harmonize society. It is the universal neurosis in that it takes individual neuroses and brings them together, not necessarily for the best (Freud 1928a).

> In my *Future of an Illusion* I was concerned much less with the deepest sources of the religious feeling than with what the common man understands by his

> religion—with the system of doctrines and promises which on the one hand explains to him the riddles of this world with enviable completeness, and, on the other, assures him that a careful Providence will watch over his life and will compensate him in a future existence for any frustrations he may suffer here. The common man cannot imagine this Providence otherwise than in the figure of an enormously exalted father. Only such a being can understand the needs of the children of men and be softened by their prayers and placated by the signs of their remorse. The whole thing is so patently infantile, so foreign to reality, that to anyone with a friendly attitude to humanity it is painful to think that the great majority of mortals will never be able to rise above this view of life. It is still more humiliating to discover how large a number of people living in to-day, who cannot but see that this religion is not tenable, nevertheless try to defend it piece by piece in a series of pitiful rearguard actions. One would like to mix among the ranks of the believers in order to meet these philosophers, who think they can rescue the God of religion by replacing him by an impersonal, shadowy and abstract principle, and to address them with the warning words: 'Thou shalt not take the name of the Lord thy God in vain!' And if some of the great men of the past acted in the same way, no appeal can be made to their example: we know why they were obliged to. (1928b, p. 21)

Best known of all those who would separate out cultural understanding from any biological underpinning is the French sociologist Emile Durkheim. His *Rules of Sociological Method* (first published in 1895) lays down categorically the independence of the sociological. He did not even want psychology explaining the social, let alone the biological! 'The group thinks, feels, and acts entirely differently from the way its members would if they were isolated. If therefore we begin by studying these members separately, we will understand nothing about what is taking place in the group. In a word, there is between psychology and sociology the same break in continuity as there is between biology and the physical and chemical sciences. Consequently every time a social phenomenon is directly explained by a psychological phenomenon, we may rest assured that the explanation is false' (1982, p. 129). What then of religion? For him, religion is a matter of practice, where an individual is taken up into a community. 'A religion is a unified system of beliefs and practices relative to sacred things, i.e. things set apart and forbidden—beliefs and practices which unite in one single moral community called a Church, all those who adhere to them' (1961, p. 62). Religion therefore has a function, bringing people together—ideas like God and so forth are secondary to this purpose. In Durkheim's terminology, it is all a question of the 'collective consciousness'.

> Thus the collective ideal which religion expresses is far from being due to a vague innate power of the individual, but it is rather at the school of collective life that the individual has learned to idealize. It is in assimilating the ideals elaborated by society that he has become capable of conceiving the ideal. It is society which, by leading him within its sphere of action, has made him acquire the need of

> raising himself above the world of experience and has at the same time furnished him with the means of conceiving another. For society has constructed this new world in constructing itself, since it is society which this expresses. Thus both with the individual and in the group, the faculty of idealizing has nothing mysterious about it. It is not a sort of luxury which a man could get along without, but a condition of his very existence. He could not be a social being, that is to say, he could not be a man, if he had not acquired it. (1961, p. 270)

In a sense, Durkheim saw religion as becoming outdated, but he saw no true successor or substitute. In the way it binds us together collectively, it has an essential role and thus far no true rival.

CONTEMPORARY EXPLANATIONS

Let us now, with this background, pick up the thread of our story, fast forwarding to the rise of evolutionary explanations of human social behaviour, specifically to contemporary thinking about possible evolutionary origins of religious beliefs. The one thing that strikes the disinterested observer is the extent to which there is a total lack of uniformity by enthusiasts. There is a variety of proposals, few of which are consistent with any other. Trying to bring some consistency or order to the diversity, we can distinguish between those who think that the causes are almost entirely biological and those who think that, although biology is an essential foundation, culture is where the real action takes place. Focusing first on the biological, we can distinguish between those who think that religion is going to be produced by natural selection because it has adaptive virtues and those (including Pascal Boyer et al.) who think that, although natural selection may be all-important in evolution, religion is something of a by-product. (I take it that this would be the position of Darwin.)

Prominent amongst those who think that religion is essentially biological in origin and that it is produced by natural selection because of its adaptive virtues is the eminent evolutionary biologist Edward O. Wilson. He argues that religion is selection-produced and of value to humans because—shades of Durkheim—of its worth as something maintaining and cohering group identity.

> The highest forms of religious practice, when examined more closely, can be seen to confer biological advantage. Above all, they congeal identity. In the midst of the chaotic and potentially disorienting experiences each person undergoes daily, religion classifies him, provides him with unquestioned membership in a group claiming great powers, and by this means gives him a driving purpose in life compatible with his self interest (1978, p. 188).

Wilson does admit that there may be something to cultural causes, but essentially he thinks that it all comes back to biology.

> Because religious practices are remote from the genes during the development of individual human beings, they may vary widely during cultural development. It is even possible for groups, such as the Shakers, to adopt conventions that reduce genetic fitness for as long as one or a few generations. But over many generations, the underlying genes will pay for their permissiveness by declining in the population as a whole. Other genes governing mechanisms that resist decline of fitness produced by cultural evolution will prevail, and the deviant practices will disappear. Thus culture relentlessly tests the controlling genes, but the most it can do is replace one set of genes with another. (p. 178)

One interesting question is whether Wilson thinks that natural selection in this case works purely for the benefit of the individual or whether he thinks (as he has been arguing strongly in other contexts) that selection here works essentially for the benefit of the group. As it happens, Wilson is rather fuzzy on the matter. (My suspicion is that today he is strongly group inclined.) Others, however, have taken strong stands on the individual/group controversy, some coming out strongly on religion as something of virtue or worth to the individual and others coming out strongly on religion as something of worth only in the group context. (Again the spirit of Durkheim hovers over this discussion for it is clear that he would incline strongly to the group perspective. For him, the individual only makes sense considered with respect to the group. That is why he turns his back on purely psychological explanations.)

Representing the former individual-centred position are the physical anthropologist Vernon Reynolds and a scholar of religion Ralph Tanner (1983). They argued strongly that religion is something which benefits and can only benefit the individual. Inventively, they rely heavily in their discussion on a distinction that evolutionists draw between so-called r selection—this being selection that works best in unstable conditions—and K selection—this being selection that works best in stable conditions. One of the major implications of the r–K division is that one expects organisms pursuing an r strategy to have lots of offspring and comparatively little parental care. This is a good strategy because, if conditions are good, then they can score strongly in the struggle for existence. On the other hand, organisms pursuing a K strategy tend to have fewer offspring and to look after them carefully. This is a good strategy in stable conditions because, while you may never have bumper years, you are under no threat of having really bad years and can even things out nicely. Reynolds and Tanner apply this kind of thinking directly to religion, arguing that those religions that encourage many offspring tend to be those religions that exist in countries where conditions are unstable; whereas those religions promoting few offspring are those religions which exist in countries with fairly stable conditions. Reynolds and Tanner point out that Catholic countries are often more variable and frequently have unstable conditions. Protestant countries, to the contrary, tend to be lands where the living conditions are significantly more stable and predictable. They argue that this is a prediction that they would expect—Catholic religious beliefs and practices tend to promote higher family sizes than do the beliefs and practices of Protestantism.

Strongly arguing for a group-selection perspective on the origin of religion is David Sloan Wilson (2002). Analysing the society that Jean Calvin founded in Geneva in the sixteenth century, he argues that we find that the religion developed there was one that put great emphasis on group cohesion: rules included 'obey parents', 'obey magistrates', 'obey pastors' and so on down the list to 'no lewdness and sex only in marriage', and 'no theft, either by violence or cunning'. This leads Wilson to write as follows:

> To summarize, the God–people relationship can be interpreted as a belief system that is designed to motivate the behaviors [examples of which are listed just above]. Those who regard religious belief as senseless superstition may need to revise their own beliefs. Those who regard supernatural agents as imaginary providers of imaginary services may have underestimated the functionality of the God–person relationship in generating real services that can be achieved only by communal effort. Those who already think about religion in functional terms may be on the right track, but they may have underestimated the sophistication of the 'motivational physiology' that goes far beyond the use of kinship terms and fear of hell. Indeed, it is hard for me to imagine a belief system better designed to motivate group-adaptive behavior for those who accept it as true. When it comes to turning a group into a societal organism, scarcely a word of Calvin's catechism is out of place. (p. 105)

Turning from those who think that religion is produced by direct selective mechanisms of advantage to humans, we find those who think that religion, although biologically caused, tends to be something of a by-product. Anthropologist Pascal Boyer, and the majority of theorists in the field of evolutionary cognitive science of religion (ECSR), incline towards this sort of thinking.

> The building of religious concepts requires mental systems and capacities that are there anyway, religious concepts or not. Religious morality uses moral intuitions, religious notions of supernatural agents recruit our intuitions about agency in general, and so on. This is why I said that religious concepts are parasitic upon other mental capacities. Our capacities to play music, paint pictures or even make sense of printed ink-patterns on a page are also parasitic in this sense. This means that we can explain how people play music, paint pictures and learn to read by examining how mental capacities are recruited by these activities. The same goes for religion. Because the concepts require all sorts of specific human capacities (an intuitive psychology, a tendency to attend to some counterintuitive concepts, as well as various social mind adaptations), we can explain religion by describing how these various mind capacities get recruited, how they contribute to the features of religion we find in so many different cultures. We do not need to assume that there is a *special* way of functioning that occurs only when processing religious thoughts. (Boyer 2002, p. 311)

See how Boyer argues that religion has no value in itself but has simply subverted or borrowed features that do have real biological value.

These are the biologists. There are also those who argue that, although biology is important, in the case of humans biology has produced culture and now we are in the situation where culture has taken over (see Leslie Newson and Peter Richerson's, and William Bainbridge's chapters in this volume). Religion therefore should be seen very much as a cultural phenomenon, with the biology in the background. One of the most popular recent theories of culture, which sees it as having been produced by biology but with biology now in the background and essentially ineffective, is the theory of memes—so-called 'memetics'—proposed by Richard Dawkins in his bestselling work, *The Selfish Gene* (1976). Dawkins argues that we should recognize cultural units akin to genes, units that he labels 'memes'. Dawkins argues that memes invade people's minds and there reproduce and are passed on to others. What is going on therefore is a selective process, with some memes being more successful than others. It is not a matter of life and death for organisms—at least not immediately—but rather life and death for the cultural units. Note that Dawkins by no means implies that memes are good for the possessors. Indeed, Dawkins is rather inclined to think that frequently—perhaps too frequently—successful memes are those very things which are not particularly good for us. Advertising clearly works on this principle.

The philosopher Daniel Dennett (2006) has taken up Dawkins theory of memes with some enthusiasm. He applies the idea directly to the evolution of religion, arguing that it is very much a parasite of the mind, just as for instance the lancet fluke is a parasite of the body. Dennett relates how this parasite first invades the brain of an ant, causing it to climb blades of grass, where it is eaten by sheep and cows. Then the fluke settles in the liver of the mammals and eventually gets excreted. Then the cycle repeats itself, as the fluke is again taken up by the ants. The fluke is of value neither to the ants nor to the cows and sheep, having adaptations which entirely serve its own ends. Religion apparently falls much in the same category, but at the cultural level.

> Does anything like this ever happen with human beings? Yes indeed. We often find human beings setting aside their personal interests, their health, their chances to have children, and devoting their entire lives to furthering the interests of an *idea* that has lodged in their brains. The Arabic word *islam* means 'submission', and every good Muslim bears witness, prays five times a day, gives alms, fasts during Ramadan, and tries to make the pilgrimage, or *hajj*, to Mecca, all on behalf of the idea of Allah, and Muhammad, the messenger of Allah. Christians and Jews do likewise, of course, devoting their lives to spreading the Word, making huge sacrifices, suffering bravely, risking their lives for an idea. So do Hindus and Buddhists. (Dennett 2006, p. 4)

People like Dawkins and Dennett obviously share with earlier thinkers like Frazer and Freud the belief that, although socially significant, religion is ultimately a set of false, and perhaps dangerous beliefs. An interesting parallel between Freud and Dennett is the extent to which they both think of religions as being not just simply bad but disease-like—in Freud's case a kind of collective neurosis, in Dennett's case an unwelcome parasite, a bit like hookworm or even worse a virus like smallpox. Note that their differences are not so much predicated on the biology versus culture divide, but on alternative types of illness or disease, purely psychological versus epidemiological. This said, if one is an enthusiast for memetics, I would have thought that Frazer's work cries out for such an analysis, with the ways in which primitive beliefs have lives of their own and chop and change according to circumstance and fashion. He is truly thinking in an evolutionary cultural way. (Less so Freud, because once the basic ideas are put in place by biology, I am not sure how much scope culture has for changing beliefs—or whether culture is needed for transmission, since we are all going to have the beliefs anyway.)

ARE THE EXPLANATIONS TRUE?

Let us change course a little now. Let us ask about the truth status of the various claims that we have just presented above. Are we talking about serious science, particularly are we talking about serious science which has a fair claim to be true? Or are we talking about speculation? In which case, are we talking about fruitful speculation which might bear further examination? Or are we really just at the level of off-the-top-of-the-head hypothesizing, which really should not be taken too seriously, at least not at this point?

One thing we can certainly say. Not every one of the claims can be true! There are far too many conflicts and disagreements to think that every proposal is well taken. If you think the biology is all-important, then obviously culture is downgraded somewhat. If you think that the key is natural selection, then suggestions that religion is just a by-product are looking in the wrong direction. If you think that individual selection is all-powerful, then clearly hypotheses which depend on group effects as the essence of evolution cannot be true. What does strike the reader vividly in going through the various accounts and suggestions is how little hard evidence is offered in particular cases; or, if such evidence is offered, how weak it turns out to be on more detailed examination.

For instance, Edward O. Wilson really does not give us much by way of firm evidential support for his claims about group solidarity. Likewise, Daniel Dennett gives little support for his belief that religion exists without regard for the good of the individual believer. He certainly makes no effort to speak to those who argue that individual believers are better people or happier people or (more importantly from a biological point of view) reproductively successful people. Among those who would argue precisely this is David Sloan Wilson. However, before we rush to praise him, although it is certainly true that he does make some effort to provide real empirical underpinning—this is what his survey of Calvinistic thought in sixteenth-century Switzerland is supposed to achieve—the sceptical reader might wonder just how detailed and relevant his evidence truly is. Perhaps he is just cherry-picking to make his case.

To be candid, the extent to which one is going to be inclined to take one or other of these various proposals seriously is going to depend less on the proposals themselves and more on one's background commitments. In other words, if you go into the discussion convinced that natural selection is all-powerful, that it applies to humans including human behaviour and thought, and that selection almost invariably works at the individual rather than the group level, then you're likely to come out feeling the proposals about the evolution of religion which fit these prior convictions are more likely plausible than any others. In this particular case, one suspects that one will respond favourably to the thinking of Vernon Reynolds and Ralph Tanner. Alternatively, if you favour the memes-eye view of evolution, then you're more likely to feel sympathy for the kind of hypothesizing in which Daniel Dennett engages. A range of other background commitments will likewise shape attitudes to the other ideas I have mentioned, including the thesis, popular within ECSR, that religion is something of a by-product.

All of this in itself is not necessarily a bad thing. It is fairly normal scientific practice to assume that a new area of investigation is going to yield to the theories and methods developed elsewhere. There is no guarantee of success but it is certainly a good heuristic strategy. If something works in one area, then see if it works in another, and don't give up until you can see definitely that the transferred application is inappropriate or unsatisfactory. However, this should be recognized for what it is. One does not have established theories but rather suggestions based on successes elsewhere. This is very much how I see present-day work on the evolution of religious beliefs. I do not think that they are stupid or inappropriate. However, they are very far from being established or even close to being established. A huge amount of work lies ahead. What one can also say is that certain assumptions and prejudices that do seem to mark present-day work will have to be addressed and eliminated. To take but one example, it is glaringly obvious—it is particularly glaringly obvious to anyone who has not been born and bred in the United States—that American scholars, addressing the evolution of religion, tend quickly to assume that what holds in their own country must hold universally. Edward O. Wilson and Daniel Dennett take rather different approaches to the whole subject, with Wilson being far more inclined to give biological evolution a major (positive) place and Dennett being far more inclined to give cultural evolution a major (negative) place. But they are united in thinking that religion is a universal phenomenon of humankind and explanations must accept this fact. To anyone who was brought up in Europe (with some exceptions like Poland), this assumption is ludicrously untrue. By and large, Europeans simply are not religious. In England, for instance, for most of the population the very suggestion that one might take seriously the major tenants of the Christian faith would be dismissed as absurd. It is not that the English are necessarily hostile to religion as that they are indifferent. People would probably be rather shocked if the Royal Family did not go to church, but in this (as in so many other things) the Royal Family is not representative of the average citizen. Clearly, therefore, any adequate analysis of religion from an evolutionary perspective must reach out beyond the narrow confines of the researchers' own backgrounds. This will not necessarily guarantee success, but one can be certain that success will not come unless these safeguards are taken. For this reason alone, it would be unwise to think of the years when cultural explanations reigned supreme as years that were wasted. A far deeper understanding of the phenomenon of religion is demanded than most of today's evolutionary thinkers appear to have. With some few exceptions—Ralph Tanner is a student of religion—the evolutionists hardly know their own society's religions, let alone those of other societies. Much of the necessary homework has been done, and it is foolish to ignore it.

IS RELIGION TRUE?

Suppose that one does grant at least some potential validity to some form of evolutionary approach to religion. Here is the final question: What implication does any of this have for the truth status of religion itself? In other words, can we say anything at all about whether or not God exists or whether or not He cares for us, given some particular evolutionary analysis of these and related beliefs? Both David Hume and Charles Darwin thought that one could separate out the causes of why people believe in religion from whether or not good reasons can be given for believing religious claims to be true. In a way, this attitude is very much akin to upholding the distinction in philosophy of science between the context of discovery and the context of justification. It does not really matter how you get your idea, what really counts is whether or not the idea is true or false. In the same way, if for instance religion is just a by-product of other things, seeing faces in the clouds or seeing spirits in wind-blown parasols, it does not mean that religion is not true. After all, if you think that God created the world according to law-like mechanisms, then religion had to be produced one way or another and ultimately any way is as good as any other. The fact that religion might have natural causal origins is no more damning to its truth status than taking mathematics seriously becomes stupid because those of our proto-ancestors who did take mathematics seriously survived and reproduced, and those of our proto-ancestors who did not take mathematics seriously did not survive and reproduce.

However there are those who, like various of our social science thinkers, believe that in fact the causal origins of religion are pertinent to its truth status. One is Edward O. Wilson, who argues that once we see that religion is no more than a product of the genes, then we start to see that religion is no more than any other adaptation without any claim to absolute validity. In his Pulitzer Prize-winning work, *On Human Nature,* Wilson writes:

> But make no mistake about the power of scientific materialism. It presents the human mind with an alternative mythology that until now has always, point for point in zones of conflict, defeated traditional religion. Its narrative form is the epic: the evolution of the universe from the big bang of fifteen million years ago through the origin of the elements and celestial bodies to the beginnings of life on earth. The evolutionary epic is mythology in the sense that the laws it adduces here and now are believed but can never be definitively proved to form a cause-and-effect continuum from physics to the social sciences, from this world to all other worlds in the visible universe, and backward through time to the beginning of the universe. Every part of existence is considered to be obedient to physical laws requiring no external control. The scientist's devotion to parsimony in explanation excludes the divine spirit and other extraneous agents. Most importantly, we have come to the crucial stage in the history of biology when religion itself is subject to the explanations of the natural sciences. As I have tried to show, sociobiology can account for the very origin of mythology by the principle of natural selection acting on the genetically evolving material structure of the human brain.
>
> If this interpretation is correct, the final decisive edge enjoyed by scientific naturalism will come from its capacity to explain traditional religion, its chief competition, as a wholly material phenomenon. Theology is not likely to survive as an independent intellectual discipline. (1978, p. 192)

It is interesting to note that, although Wilson does not want to give religion any ultimate truth value, unlike some (for instance Richard Dawkins and Daniel Dennett) although like others (Durkheim notably) Wilson thinks that religion has an important social value. Therefore, rather as in the *Republic* Plato argues for the virtues of the noble lie, the belief that humans are essentially made up of gold, silver, or bronze, so Wilson argues for the virtues of his noble lie, the belief that religion is true. Where he would take one step more is in arguing that we require a new religion for the new age. In recent years, he has been promoting just such a vision, one which pushes a kind of humanism based on a progressive reading of the evolutionary process. (Wilson is a major enthusiast for the ideas of the nineteenth-century evolutionist Herbert Spencer. Much that Wilson advocates today can be found in the writings of that quintessential Victorian.)

My suspicion is that basically Hume and Darwin are right. If the belief in mathematics can be explained naturally, it does not follow that mathematics is not necessarily true. Likewise, if the belief in religious systems can be explained naturally, it does not follow that the religious systems are not necessarily true. This does not mean that they are true, but rather that a naturalistic explanation does not show them to be false. However, I suspect some qualifications need to be made here. Having full regard for the nature of the claims that religion makes, I suspect that certain, possible, natural origins of religion would sit uncomfortably for the true believer. We know that the evolutionary process causes a great deal of pain and suffering. Hence, if we are religious believers, we have to accept that the God that we believe was creator allowed this pain and suffering as part of the creative process. It may well be that God is all-loving and all-powerful, but still we must allow that God almost positively endorses the agonies that were required for organisms to exist as they do today—organisms including human organisms.

I am just making this point and not intending here that it be taken as a refutation of Christianity in particular or religion in general. As is well known, religious people have many arguments to counter the problem of evil. What I am saying is that causal origins are not entirely irrelevant to the contents of the beliefs, and that particular causes and particular contents might sit uneasily together and require a certain amount of clarification and argumentative justification. Let me push this point. I could imagine origins that it would be very, very difficult indeed to reconcile with Christian belief. Suppose, to take a rather unpleasant example, one could show that religious beliefs come because and only because of instances of sexual assault. It is only after somebody has been violated sexually that, either in their own minds or the minds of others, religious beliefs start to spring up and can spread. Please note, I am not saying that this is in any sense plausible but simply proposing it as a hypothetical example. But suppose that it is true. One therefore seems to be in the position of saying that religious beliefs come about through, and only through, really vile evil acts. Does one really want to say that this is compatible with Christian belief?

Or if you want a slightly more plausible example, take the Holocaust. The story of Job shows very well that, when people are suffering, they tend to turn to God rather than away from Him. The Holocaust involved extreme suffering. Suppose then that sociologists and others are able to show that religious beliefs persist only when really dreadful events like the Holocaust occur. Is this really something that one expects of a loving, powerful deity? Without answering this question fully one way or the other, I think at least it must be allowed that it does pose a major issue to be discussed. Hence, what I am saying is that, although logically religious beliefs can be held no matter what the causal origins, I am not convinced that the causal origins are entirely irrelevant to the truth status of the religious beliefs. The point I am making is one that would be strongly endorsed by Frazer. That there are echoes in other societies of the story of the Incarnation and the Atonement—the coming of the living god and his death on the cross for our sins, followed by his coming again to life and thus making possible our salvation—do not in themselves make the story false. But you do start to wonder if our story is but one among many and that we believe it and not the others simply because it is ours, it is the one that we learnt as children. If we reject the others then candour demands that we reject ours. Even if it is true, we have no good reason to think it true. The causal understanding of religion makes the epistemological justification shaky to the point of absurdity.

This brings us to the final question of whether or not any of the proposed causal explanations sketched earlier in this paper fall anywhere close to the kinds of repellent, causal processes that I have just been discussing. Obviously most do not. When Edward O. Wilson argues that religion promotes group cohesiveness, one is about as far from sexual assault or the Holocaust as one can possibly imagine. In fact, if Wilson is right, far from the Christian accepting Wilson's own conclusion that Christianity is therefore false, the Christian might well think that we have the sort of warm and friendly origins that one would expect of his or her God. However, perhaps expectedly given that Daniel Dennett is a violent critic of any and all religious beliefs, one might think that his parasite-based explanation does cast doubt on the truth status of religion. One suspects that this was very much his intention. If religion is basically the cultural equivalent to the lancet fluke, then it is hard to think that it is something of great value and worth to humankind.

It is surely obvious that if one is a believer and if one thinks there is something to Dennett's scenario, then one should be addressing the issue. One way to do this obviously would be to show that Dennett's claims are empirically false. Independently above, I have already suggested that his claim is not obviously well taken, given the extent to which they are so American-centric. In the same vein, one might attack the whole theory of memetics generally. There are many inclined to do this, given the theory's inability to offer any new or fruitful predictions (see Ruse 2009). Another way to counter Dennett might be to suggest that even if memetics be true, the parasite analogy is not necessarily the best one when thinking about religion. And even if the parasite analogy is well taken, then one should recognize that sometimes parasites are of value rather than otherwise. For instance, when termites shed their outer coverings (which they do several times during their lives), they lose the parasites necessary to digest the cellulose on which they live. They therefore needed to pick up the parasites again from other species members. So perhaps religion could be likened to the sort of parasite which, far from being unpleasant and dangerous like the lancet fluke, is of important social and physical value.

CONCLUSION

There is today a range of suggestions about the possible evolution of religious states. Not all of them are particularly well taken, but if one takes seriously evolutionary biology then, at the least, we should allow the possibility that one or more of the proposals will develop into deeper explanations of our religious beliefs. As noted, it would be foolish to think that this should be done in opposition to the work of anthropologists and sociologists and others. If any real advances are to be made, it is surely going to be a joint effort rather than something done in ignorance of the findings and theorizing of others. Whether or not any of this has implications for the truth status of religion is another matter. As a general rule, probably it does not. However, one should be wary of absolute judgments at this point. Depending on the nature of the explanations, it may be that the religious believer will need to rethink and perhaps to defend and even amend the beliefs that he or she holds dear.

ACKNOWLEDGEMENTS

I am much in debt to my colleague Matthew Day for discussions on the topic of this essay and for the insight about the parallels between the thinking of Freud and Dennett.

REFERENCES

Boyer, P. 2002. *Religion Explained: The Evolutionary Origins of Religious Thought*. New York: Basic Books.

Darwin, C. 1871. *The Descent of Man, and Selection in Relation to Sex*. London: John Murray.

Dawkins, R. 1976. *The Selfish Gene*. Oxford: Oxford University Press.

Dennett, D. C. 2006. *Breaking the Spell: Religion as a Natural Phenomenon*. New York: Viking.

Durkheim, E. 1961. *The Elementary Forms of the Religious Life*. New York: Collier.

——. [1895] 1982. *The Rules of Sociological Method and Selected Texts on Sociology and its Method*. Ed. S. Lukes. New York: Macmillan.

Frazer, J. G. 1890. *The Golden Bough: A Study in Magic and Religion*. London: Macmillan.

——. 1911. *The Golden Bough: A Study in Magic and Religion. Volume 4. The Dying God*. London: Macmillan.

Freud, S. 1928a. *Civilization and its Discontents*. London: Hogarth Press.

——. 1928b. *The Future of an Illusion*. New York: Horace Liveright and the Institute of Psychoanalysis.

Hume, D. [1757] 1963. A natural history of religion. In *Hume on Religion*. Ed. R. Wollheim, London: Fontana, 40–1.

Reynolds, V. and R. Tanner. 1983. *The Biology of Religion*. London: Longman.

Ruse, M. 1979. *Sociobiology: Sense or Nonsense?* Dordrecht, Holland: Reidel.

——. 1996. *Monad to Man: The Concept of Progress in Evolutionary Biology*. Cambridge, MA: Harvard University Press.

——. 2008. *Charles Darwin*. Oxford: Blackwell.
——. Ed. 2009. *Philosophy after Darwin: Classic and Contemporary Readings*. Princeton: Princeton University Press.
——. Ed. 2013. *The Cambridge Encyclopedia of Darwin and Evolution*. Cambridge: Cambridge University Press.
Sulloway, F. 1979. *Freud: Biologist of the Mind*. New York, N.Y.: Basic Books.
Wilson, D. S. 2002. *Darwin's Cathedral*. Chicago: University of Chicago Press.
Wilson, E. O. 1978. *On Human Nature*. Cambridge, MA: Harvard University Press.

4

The Evolution of Divine and Human Minds: Evolutionary Psychology, the Cognitive Study of Religion and Theism

AKU VISALA

It is often assumed that cognitive and evolutionary accounts of religion make religious truth-claims problematic, or at least more improbable than they would be in the absence of explanations of the psychological and biological causes of religious behaviour. Nevertheless, several recent papers have argued that such explanations need not have fatal consequences for religion. In what follows, I shall provide an overview of these debates. I will also look at the relationship between cognitive and evolutionary explanations of religion, especially those of Evolutionary Psychology (EP), within these debates. Typically, in these debates, the distinction between evolutionary and cognitive explanations is unclear or not made at all. What interests me is whether the debates would be any different if we were to draw this distinction carefully.

EVOLUTIONARY EXPLANATIONS AND COGNITIVE EXPLANATIONS

In its standard usage, the term 'cognitive science of religion' (CSR) refers to an interdisciplinary research programme that seeks to explain the prevalence and persistence of religious beliefs and behaviours by invoking natural cognitive faculties and behavioural tendencies. For the most part, discussions about the religious impact of new naturalistic theories of religion have focussed on CSR and its results. For clarity's sake, we must now ask how CSR relates to evolutionary and biological explanations of religion.

We can begin by looking at the explanations in general terms. Explanations have three components:

1. The thing (event, tendency, etc.) that is being explained. Let us call this the *target* of the explanation.
2. The thing (event, tendency, etc.) that provides the explanation. Let us call this the *explainer*.
3. The way in which the explainer explains the target.

In the case of religion, the target is not a simple object, but a set of multiple, complex phenomena conceptualized in a certain way. As many representatives of CSR also admit, religion as a category is too broad and vague to be explained as a whole. Boyer suggests that we should not presume that 'religion' is a homogeneous category at all, but rather a collection of different kinds of activities, events, concepts, and representations. As there can be no single explanation for individual representations of things as diverse as rituals, god-concepts, religious specialists, misfortune, and so on, there cannot be a single 'theory of religion' (Boyer 1994). Thus, in order to make progress in explaining religion, we need to slice the phenomenon into smaller, better-defined bits.

Components (2) and (3) sometimes overlap but it is useful to distinguish them conceptually. Traditional anthropologists and sociologists of religion have suggested explainers such as experiences of the sacred and social structures. In CSR, the explainers are the cognitive mechanisms underpinning all transmission of cultural information, including religious information. By their relatively invariant operations across different cultures, they drive cultural and religious information towards certain forms; they are the factors that drive the selection process of cultural evolution.

This leads naturally to the question of how the explainer explains the target. What is being explained by CSR is a general human disposition to prefer religious concepts and information by invoking certain features in the operation of panhuman cognitive systems. In this sense, CSR explanations resemble psychological explanations: they do not refer to single events as targets and explainers, but rather to mechanisms with functioning parts that produce the target disposition (Cummins 2000; Visala 2011).

Now that we have a basic idea of what cognitive explanations of religion are like, let us focus on evolutionary explanations of religion. First of all, we must distinguish evolutionary analogies from evolutionary explanations. This distinction between analogical and literal uses of evolutionary biology in explanations has been made elsewhere (Haines 2007). We can talk about evolutionary analogies when cultural evolution is conceptualized and explained in terms analogous to those of biological evolution. In the case of the CSR explanation described earlier, biological evolution is used in an analogous sense. What is being explained is the cultural fitness of certain types of representations. This is cultural, not biological evolution. Theories and concepts of biological evolution are used to model and explain culture.

Evolutionary explanations come in many varieties: we can seek to explain, for instance, why certain traits are especially prevalent in some populations or how a trait develops in the lifetime of an individual organism. Sometimes explanations of the former sort are referred to as ultimate explanations and the latter as proximate explanations. In this sense, ultimate explanations explain the genetic features of populations in terms of selection pressures and adaptations, whereas proximate explanations explain how the genetic 'programme' is decoded in the development of individual organisms. To give an ultimate evolutionary explanation of religion, then, would be to explain what selection pressures lead to the formation of 'religious traits' in human groups. Many evolutionary explanations of religion suggest that religious thinking and behaviour have adaptive value for individuals or groups and this explains why religious beliefs and behaviour have become so common in human populations (Sosis and Alcorta 2003; Bulbulia 2009; see also Ilkka Pyysiäinen's article in this volume).

Those who seek to provide evolutionary and cognitive explanations of religion are divided as to what role these explanations play in the big picture. This is sometimes known as the by-product/adaptation divide. Those who emphasize religion's adaptive functions seek to give a larger role to evolutionary explanations, whereas those who focus on cognitive mechanisms in explaining religion think that religion is more a by-product of normal cognition than an adaptation produced by natural selection.

At this point it needs to be said that the adaptation/by-product divide is not as simple as some CSR writers, especially Atran and Boyer, want to make it. These convictions should not be seen as mutually exclusive or rigid alternatives. This is mainly because of the problematic notion of 'adaptation' itself. As Jeffrey Schloss points out, the claim that religion is an adaptation is vague (Schloss 2009). Consider the following options. It might very well be that some religious phenomena are adaptations and others not. Furthermore, it might turn out that some religious behaviours were originally adaptations to some specific set of circumstances, but are no longer adaptive. Finally, it might be the case that some religious behaviours came about as by-products, but then turned out to be favoured by natural selection in some environments (these are often called exaptations). The point that I am trying to make here is that we do not have enough data to decide between these alternatives. Nor are we compelled to choose.

There is also a second reason to avoid rigid oppositions, namely, that evolutionary hypotheses seem to be set at different levels of explanation than cognitive ones. Pyysiäinen also makes this point in his chapter. For instance, we could regard the CSR standard model as giving us proximate mechanisms of religion whereas an evolutionary explanation gives us ultimate explanations. It might be that the CSR standard model can be combined with and subsumed under a larger Darwinian explanation. In other words, religion could be a by-product on the psychological level but that does not give us much in the way of clues as to how it might be understood at the biological level.

Even if it is the case that evolutionary and cognitive explanations of religion do not mix at all, it does not mean that evolutionary explanations have nothing to do with cognitive explanations. In the standard model of Evolutionary Psychology, the idea is that our evolutionary history explains the way our cognition works and cognition explains how our culture works. In the case of religion, human cognition explains why religion is what it is, whereas an evolutionary explanation can be given for the architecture of our cognition. Although many CSR writers emphasize the evolutionary background of the human mind's cognitive architecture, it is not the case that all writers in the CSR field have an equal amount of interest in this topic. Atran (2002) and Boyer put forward large numbers of evolutionary hypotheses whereas Barrett (among others) seldom poses evolutionary questions.

HOW TO THINK ABOUT RELIGIOUS RELEVANCE

In order to assess the religious relevance of naturalistic approaches to religion, we must have a basic idea of what it means for some scientific theory to have religious implications. In what follows, then, I want briefly to explain the notion of religious relevance.

First of all, we should avoid discussing the relationship between naturalistic explanations of religion and religious truth-claims in abstract terms. There is no such thing as 'religion' in any general sense. Instead, we have a wide variety of practices, beliefs, and world-views. So we should always focus on some particular theological theory or idea, not just on generic 'religion'. Since most of the participants of these debates have been either Christians or other theists, I will be focussing on theism, that is, the claim that there exists an unsurpassable being who is the creator of everything outside itself, necessary, omnipotent, omniscient, and perfectly good. Of course, Christian theism includes many other claims as well. Even if cognitive and evolutionary explanations of religion might not have religious relevance for theism *per se*, they might be relevant for some specific theological claims about the nature of human beings, morality, freedom, sin, and so on. Because space is limited, I will focus my attention on bare theism.

Anyone who wants to say something about the compatibility (or incompatibility) of cognitive and evolutionary explanations of religion and religious claims needs to be very careful how they frame the question. Let us define religious relevance in the following way:

> *Religious relevance*: a scientific theory T is relevant for a theological theory TH if the truth of T makes the truth of TH either more plausible or implausible.

I want to flesh out the expression 'more plausible or implausible' within a simplified Bayesian framework. I propose that plausibility and implausibility can be viewed as relative to old evidence upon which the epistemic subject draws and new evidence that the subject learns. In these terms, our epistemic situation is as follows. We hold a reasoned view (hopefully) about the degree that evidence supports theism. Now, new evidence emerges, namely, cognitive and evolutionary theories of religion. Our question then is whether we should adjust our view of the plausibility of theism up or down on the basis of this new evidence. If the plausibility of theism goes up, then cognitive and evolutionary theories of religion are positively relevant for this religious claim, and if it goes down, the relevance is negative.

Finally, there is the question of when evidence supports or undermines a given proposition. Consider the following example: if there were a theistic God, we would expect to find complex biological organisms capable of life, thinking, and morality in the universe. Thus, any evidence for these facts supports theism and makes theism more likely to be true than without this evidence. This does not mean that we should, all evidence considered, become theists. The existence of complex biological organisms is also evidence for many other hypotheses as well, not just theism. Standard evolutionary naturalism, for instance, claims to explain the emergence of complex biological life without invoking divine intervention. If this is correct, then the fact of complex biological life does not make theism more plausible than naturalism, although it makes theism more plausible than it would be if there were no life at all in the universe.

The benefit of the aforementioned framework is that it allows us to discuss the religious relevance of cognitive and evolutionary explanation without taking positions about the plausibility (or implausibility) of theism (or naturalism) given the total evidence. In other words, we do not have to adopt theism, atheism, or naturalism or debate the plausibility or implausibility of these views in any general sense. Thus, it might very well be the case that cognitive and evolutionary explanations of religion are negatively relevant for theism, but we could be theists anyway, because of the plausibility that theism has, given the total evidence. Conversely, it might very well be the case that cognitive and evolutionary explanations of religion are positively relevant for theism or some related claims, but we might still be justified in holding onto our atheism (again, given the total evidence).

ARGUMENTS AGAINST THEISM

The arguments in this section are discussed in more detail by Leech and Visala (2012). Only very seldom does one see arguments against theism from cognitive and evolutionary explanations of religion spelled out in any rigorous way. Often negative implications are simply assumed rather than explicitly argued for. Archaeologist Steven Mithen is a good example of this. Mithen begins by arguing that religiosity is a ubiquitous feature of human societies. He then says that it can be explained in two ways: (1) either there exists a supernatural realm inhabited by a supernatural agent or several, or (2) the human mind is itself prone to believe in the supernatural and the 'on-going activity of the universe and life are explained by entirely natural processes' (Mithen 2009, p. 11). He concludes that 'religious thought is uniquely associated with *Homo sapiens* and arose as a consequence of cognitive fluidity, which was in turn a consequence of the origin of language. In this regard, there appears to be no need to invoke a moment of divine intervention that initiated the start of a revelation. For me, therefore, there is no supernatural, no God to be revealed' (Mithen 2009, p. 27).

The argument seems to be that because we now have a causal explanation for the emergence of religious ideas, we can exclude the possibility of supernaturalism being true in general. A deductive version of the argument would be something like this:

1. If we have a complete or sufficiently complete causal explanation of how belief in God came about, without any reference to the existence of God, then God does not exist.
2. CSR can provide a complete or sufficiently complete causal explanation of how belief in God came about without any reference to the existence of God.
3. Therefore, God does not exist.

The proponent of the argument needs to give some reasons for both 1 and 2. Mithen gives no reasons for 1, but he does give reasons for 2. He claims that, by combining both cognitive and evolutionary explanations of religion, we can get a reasonably good naturalistic explanation for why different kinds of religions persist and how they originated.

Now, why should the theist (or any other opponent of the argument) accept it? First of all, there seems to be very little reason for 1. We cannot simply infer God's non-existence from the mere fact that there exists a causal explanation for why we believe in God that does not involve God. The causal explanation and God's existence could be compatible; that is, there is no logical contradiction between these two states of affairs. It is a fallacy to assume that simply by exposing the causal history of a set of beliefs (even if complete and true) we get to the falsity of the propositions involved. This is because the truth (or falsity) of a proposition has no necessary relationship to the causes that led people to believe it.

Suppose that there is a group of people who believe in the essential difference of men and women. They believe that there is a distinctive essence to each sex. Now, a psychologist comes along and says that humans in general have a tendency to think in terms of essences—this is normally called essentialism. Given the essentialist tendency, it is easy for humans to believe in the essential nature of sex difference. Thus, the psychologist has given a causal explanation for believing in sex difference without making any reference to the facts themselves, that is, the actual nature of the differences between the sexes. But does it follow that the proposition 'there is a male essence and a female essence' is false? Of course not. It might very well be the case that there is a male essence and a female essence, even when we have a psychological tendency (no matter how strong) to believe in their existence. Similar reasoning can be applied to theism. The fact that we have a causal explanation for our belief in the existence of God without invoking God, does not make the proposition 'there is a God' false. According to philosopher John Mackie, 'Even an adequate, unified, natural history which incorporated all these factors would not in itself amount to disproof of theism. As William James and many others have insisted, no account of the origin of a belief can settle the question whether that belief is true or not' (Mackie 1982).

The second point I want to make is that claim 2 is also vulnerable to critique. It seems that we have good reasons to think that the CSR causal account of religious belief is *not* one that would allow us to exclude the possibility of God being directly involved in the process. The critic can proceed in many ways. First, the critic can argue that we do not know yet whether the current cognitive and evolutionary explanations of religion are the best explanations for the emergence of religion. The critic can grant that perhaps the cognitive and evolutionary factors that current theories invoke to explain religion are necessary, but not sufficient conditions for the emergence of religion (or at least we do not know that they are).

The critic can also invoke the nature of cognitive and evolutionary explanations themselves. What these explanations do is explain general behavioural and belief-forming tendencies in human groups. They do not determine individual belief-formation or content-specific religious beliefs in religious traditions. Consider, for instance, the doctrine of the Trinity. Can any cognitive or evolutionary explanation predict the emergence of this group of beliefs? It seems quite unlikely. Cognitive explanations can predict that human communities will believe in supernatural agents, organize around them and link morality with these agents. Our cognitive mechanisms surely shape the way in which Christians think about God, but do we have any evidence that they would ultimately cause them to develop the doctrine of the Trinity? Further,

CSR cannot explain why given individuals hold particular theistic beliefs. Particular theistic beliefs (along with all other religious beliefs) are presumably supported by the cross-cultural human intuitive biases specified by CSR. However, they do not all by themselves produce particular theistic beliefs. In other words, it would be a misconception to suppose that CSR explanations specify all of the causal pathways through which people come to hold theistic beliefs. Some anthropologists have attacked cognitive and evolutionary explanations of religion for just this reason: they do not have much to say about particular cultures (Laidlaw 2007; Day 2007). While I think that the point these authors make does have some validity, it is also too simple. Some cognitive explanations have turned out to be surprisingly effective in explaining specific cultural practices (Cohen 2007).

In short, the critic can respond to 2 by claiming that the CSR explanatory model underdetermines how humans acquire and transmit religious beliefs and behaviours. It follows from this that simply on the basis of the CSR results one cannot exclude the possibility of other factors (including supernatural factors) being involved in the causal mix (Visala 2011).

Now, let us back up a little bit. Perhaps I misunderstood Mithen's argument when I constructed it as a deductive argument for atheism. Perhaps the argument is not really for atheism at all, but a more modest parsimony argument for the conclusion that supernatural factors are superfluous in explaining religion. In this case, the argument would be something like the following.

4. We should not postulate more factors to explain religion than we need in order to adequately explain it.
5. We can explain religion in terms of cognitive and biological factors.
6. Therefore, we should not postulate supernatural entities or causes to explain religion.

Claim 4 is derived from a very basic notion of parsimony: only postulate the entities you need to explain something, no more. Premise 5 is basically supported by the same reasons as premise 2 and is therefore subject to the same criticism. The opponent of the argument can, thus, reject 5. But let us suppose that 5 turned out to be true. What then? Should the opponent of the argument try to argue against 4? I do not think so: the parsimony principle has good reasons behind it—reasons which are, for the most part, accepted by theists as well.

If the theist grants 5 and the argument goes through, does 6 provide problems for him? It might make some specifically Christian claims about theological anthropology or some aspects of religious practice somewhat problematic, but it seems that it would not do much to the case for theism. None of the standard arguments of natural theology (e.g. cosmological, teleological, or ontological arguments) invoke God's miraculous or supernatural action in religions as evidence for God's existence. Provided that particular historical episodes of special revelation are not undermined by cognitive explanations (e.g. life and 'afterlife' of Jesus), accepting 6 seems to do very little for the plausibility of the case for Christian theism. This might actually happen if cognitive explanations explain the particular historical facts far better than any supernatural explanation.

Perhaps we could understand the argument against theism in inductive terms.

7. If God exists, it is probable that the causes of our belief in his existence are supernatural rather than natural.
8. Given the results of CSR, it seems probable that the causes of our belief in the existence of God are more natural than supernatural.
9. Therefore, CSR makes the existence of God more unlikely than not.

There are various avenues open for the opponent of the argument. The first one is simply to bite the bullet and admit that CSR is indeed evidence against theism, but at the same time maintain that given the total evidence, theism is still more likely than its alternatives.

But let us say that, for whatever reason, the opponent does not want to take the hit straight to the chin and wants to challenge the argument. One option is to argue against 8 along the lines introduced earlier in conjunction with claim 2. In short, one could claim that CSR has nothing to say about individual belief-formation, personal religious experience, or revelation, for example.

Premise 7 is usually just assumed but seldom spelled out. The assumption is that, if there is a God of classical theism, God would, with high probability, make his existence known to us and enter into a relationship with humans in a way that would be clearly or mostly supernatural in a way detectable to science. Of course, not even the hardened supernaturalist thinks that there are no natural causes at work in belief in God. The proponent of the argument would have to insist that there are many more natural causes at work than theism would have given us reason to expect. Now, the question is whether this claim can be reasonably backed up.

It seems that God (who is omnipotent) would certainly be capable of using natural causes, such as our natural cognitive systems and cultural evolution, to supplement supernatural influence (such as divine action of the Holy Spirit) to a large extent (van Inwagen, 2009). Furthermore, mainstream Christian theology has been quite 'anti-supernaturalistic' in such matters. Although faith is understood in many ways to be a supernatural gift, the pathways through which it is acquired do not seem to be very supernatural: reading the Bible, listening to sermons and participating in worship and Eucharist, for example. It is said that these are *media gratiae*, mediums of grace, through which faith is given to a Christian. Given this, it seems to me that mainstream Christian theology expects the direct causes of individual religious faith to be more natural than supernatural. So, it seems that the Christian has no pressing reason to accept 7.

ARGUMENTS AGAINST THE RATIONALITY OF THEISM

The arguments in this section are dealt with elsewhere (Visala 2011; Leech and Visala 2011). Many arguments have been made against the rationality of believing theism on the basis of what we know now about the causes of religious belief. These arguments attack the rationality (or justification) of religious belief, not its truth. The claim is, basically, that the pathways through which our belief in God comes about are not reliable or trustworthy. Unreliability arguments seem at least *prima facie* plausible: if we can demonstrate that belief in God is completely naturally caused, that is, not in any direct way connected to God, then it seems that our belief that God exists is not properly truth-sensitive.

In the philosophical literature, arguments of this kind are sometimes called debunking arguments or evolutionary debunking arguments (Barrett 2007; Barrett, Leech, and Visala 2010; Clark and Rabinowitz 2011; Leech and Visala 2011, 2012; Murray 2007, 2009; Murray and Goldberg 2009; Schloss and Murray 2009). Philosopher Guy Kahane provides a schematic version of the following argument (Kahane 2011):

> S's belief that p is explained by X.
> X is a process that does not track the truth.
> Therefore, S's belief that p is not justified.

If we fill in the details, we get the following argument:

10. S's belief that there is a God is explained by CSR (that is, by S's intuitive cognitive mechanisms).
11. The cognitive mechanisms that cause belief in the existence of God are not truth tracking with respect to the existence or non-existence of God.
12. Therefore, S's belief that there is a God is not justified.

Let us examine premise 11 first. At least the following reasons can be given in its favour. First, it could be argued that a proper kind of causal connection has to exist between a belief and its target for the belief to be justified. Such a link does not exist in the case of theism, because if God did not exist, people would be theists anyway. Second, some specific cognitive mechanism postulated by the CSR could be singled out as not being a reliable source of beliefs. Take what Barrett calls the hypersensitive agency detection device (HADD) as an example. It could be argued, that since HADD produces so many false positives (false beliefs), it is not to be taken as a reliable source of beliefs about God. Third, it could be claimed that the mechanisms that CSR identifies produce all sorts of wildly diverse beliefs about God and gods. From this actual diversity of beliefs we can then infer that we should not take such systems as reliable sources of beliefs about God.

The critics of this premise have responded to the claims by arguing that they fall short of showing that the cognitive mechanisms that produce the belief in the existence of God are unreliable. In responding to the first issue of the purported lack of a proper causal connection, Michael Murray points out that, as far as we know, God might be active further back in the causal nexus and set up the initial physical conditions of the universe and guide evolution in such a way that we would develop capacities to form beliefs about him. In this case, there would be no need for God to intervene by special divine action at every step of religious belief formation and thus the completeness of the CSR account is preserved (Murray 2009; Murray and Goldberg 2009). Another response would be simply to reject the need for a causal connection between a belief and its target altogether. The causal connection view has many problems of its own, such as the difficulty of explaining how we can have justified beliefs about abstract, mathematical truths or universal generalizations that do not seem causally active.

Murray and others have also responded to the purported unreliability of HADD. Their argument is that the unreliability of HADD in the context of religion cannot be demonstrated without first assuming that there is no God. Most of the time our agency detection is correct in attributing agency. Simply on the basis of CSR, as far as we know, it could also be reliable in the case of God. In addition, it could be insisted that since even CSR writers themselves disagree as to what the explanatory scope and function of HADD actually is, its role in explaining individual instances of belief in God might actually be quite small (Barrett 2004).

Thirdly, the theist can argue that the diversity of religious beliefs does not really count against the reliability of the cognitive mechanisms that produce them because the diversity is actually a product of the cultural context that interprets the basic outputs of the mechanisms. The mechanisms themselves only produce certain intuitions to which different kinds of elaborated cultural constructions can then attach themselves. Thus, the diversity of religious beliefs does not give us enough reason to think that the underlying systems are unreliable sources of belief in God.

Premise 10 could also be challenged along the lines of claims 2 and 8. If these points are correct, then any given individual's belief in God is underdetermined by CSR. This is where reasons, arguments, testimony, religious practice, or even direct supernatural influence can come into play. If this is the case, then even if CSR turns out to be true and we could demonstrate that the cognitive mechanisms that CSR reveals are unreliable, the theist might still be justified in believing in God on the basis of these other reasons (Leech and Visala 2011).

POSITIVE RELIGIOUS RELEVANCE

Some have argued that the results of CSR might have positive religious relevance as well, namely, that some of the CSR results support theism directly or some more specific doctrine.

First of all, it is clear that the results of CSR have positive relevance for theism in at least one obvious sense: the naturalness of religion supports the notion that humans have a natural capacity to think about God. Although certain twentieth-century theologians have claimed that there is nothing in human nature that prepares us for special revelation, the standard view has been (and is) that humans are, by nature, beings that are open towards the non-natural, towards God. According to this view, humans have the required natural capacities to think about God and these capacities are there because God intended them to be. Given this, theism will predict that a general 'religiosity' or openness towards God is a ubiquitous state of affairs. Therefore, the evidence that CSR provides of the naturalness of religion gives support to theism.

Some have also claimed that the naturalness of religion hypothesis, if true, also supports theism in an indirect way: if religious thinking is natural for humans, then the cognitive mechanisms involved are the same as in other types of thinking. Further, if we normally take these cognitive mechanisms to be reliable, they seem *prima facie* reliable in the case of religious thinking as well (Gregersen 2006; Barrett 2004). Without a separate argument for their unreliability in religious matters, we cannot simply reject their outputs as unreliable.

Even if this point about the *prima facie* rationality of theistic belief is correct, it does not give us a reason to think that theism is true—it does not make theism more probable than it would be without it. What it does do, if correct, is make the basic cognitive mechanisms *prima facie* reliable sources for beliefs about God. Whether they actually produce justified theistic belief depends of course on many factors, including, for instance, whether God exists and chooses to communicate knowledge about himself, and whether there are conclusive arguments against the *prima facie* justification and so on.

This idea of prima facie justification of theistic belief is sometimes connected with the idea of what some theological traditions have called *sensus divinitatis*, the 'God-sense' (Barrett 2009; Clark and Barrett 2010; Plantinga 2000). According to traditional Calvinist theology, the *sensus divinitatis* is the source of our basic, non-inferential beliefs about the nature and existence of God and is providentially guided by God. *Sensus divinitatis* is also a mechanism that all properly functioning human beings have even if they do not recognize that they have it. If theism really predicts the existence of such a mechanism and if it resembles the mechanisms that CSR describes, then it seems that the results of CSR are evidence for this view (Leech and Visala 2012).

EVOLUTIONARY EXPLANATIONS, EVOLUTIONARY PSYCHOLOGY, AND RELIGIOUS RELEVANCE

In this last section, I will discuss the question of whether grounding CSR explicitly in Evolutionary Psychology (EP) makes any difference to the religious relevance it has for theism. As I mentioned before, the nature and depth of the link between CSR and EP is contested by people working in the area.

First of all, a few words about EP. I have so far used the expression with capital letters. There is a reason for this. We must realize that there is no single 'evolutionary psychology'. Rather we should distinguish evolutionary psychology (EP), the paradigm (with capital letters), from the field of evolutionary psychology, as David Buller has suggested (Buller 2005). Evolutionary psychology as a field of research is a broad and loosely connected group of diverse disciplines, such as behavioural ecology, human ethology, and evolutionary anthropology. In this sense, evolutionary psychology is a general term for approaches that use contemporary evolutionary theory as a background for psychological, anthropological, or neuroscientific theorizing. However, Evolutionary Psychology (with capital letters) is used to refer to a specific paradigm advocated by a group of scholars with relatively well-defined theories and background assumptions. The main persons in this group are psychologists Steven Pinker, David Buss, and Leda Cosmides, and anthropologist John Tooby. The basic ideas are that (1) the human mind is a massively modular system (2) whose modules were created by natural selection (3) to solve adaptive problems in our ancestral (Pleistocene) environment. Today (4) these once adaptive modules perform many tasks that they were not selected for and (5) form the basis of our modern cultural forms. Most people in EP believe EP to have resources by which sociocultural and behavioural sciences can be integrated with biological sciences (Barkow, Cosmides, and Tooby 1992; Buss 1999; Pinker 1997).

Let us say that CSR is linked to EP in such a way that CSR accepts the basic assumptions of EP (Boyer 2001). Culturally transmitted behaviours and practices would then be (as many CSR writers already suggest) evolutionary by-products rather than adaptations. Would this make any difference with respect to the arguments discussed in earlier sections? I do not see how it would. The empirical premises of these arguments contain nothing that would be changed by acceptance of EP.

But perhaps EP has religious relevance by itself. This question is too big to be addressed here in any detail. Let me just say this: EP does not make any difference to whatever religious relevance evolutionary biology in general has for theism. If the theist is willing to accept that humans, animals, and other living beings have evolved, then EP only fills in the specific details about the evolution of homo sapiens. In this sense, the religious relevance of EP for theism collapses into the religious relevance of biological evolution in general. Of course, this conclusion does not in any way exclude the possibility of biological evolution being strongly negatively relevant for theism.

But perhaps things are not as simple as this. Some have thought that if religion has an evolutionary explanation in terms of adaptation rather than being simply an evolutionary by-product, this would somehow be better for theism (van Slyke 2011). Religion would then be, or so the argument goes, something that produces good effects in its adherents and this is something that we would expect if theism were true. The crucial assumption is, of course, that if God exists, we would expect that being religious produces traits that make religious people more likely to survive and reproduce.

Why should we think that? First of all, the fact that religious beliefs and behaviours might have adaptive value in some way does not necessarily mean that religion provides those benefits to its adherents now, in the cultural context in which they find themselves. Adaptations are relative to the environment, and when the environment changes some traits that had adaptive value before might become maladaptive. Furthermore, when a biologist talks about a trait being beneficial, it simply means that the trait in question makes the organism more likely to survive and reproduce. I suspect that most people who think that, given theism, religion must be an adaptation have confused biological benefits with the flourishing of individuals. Even if religious behaviours and beliefs were adaptations, this would not mean that engaging in them would make humans more likely to flourish in the environment in which they find themselves.

Secondly, most contemporary evolutionary explanations of religion claim that the benefits of religion favour pro-social behaviour and in-group solidarity. For example, religious commitment might function as a costly signal of trustworthiness (Sosis and Alcorta 2003) or remove the need of external policing as the supernatural punishment theory supposes (Johnson and Bering 2009). In these cases, religious commitment is adaptive in the sense that it makes religious people better co-operators. But this readiness for cooperation has its price: the increase of out-group suspicion and hostility. By strengthening in-group commitment, the likelihood of out-group hostility grows. It is difficult to see why the theist would think that out-group hostility is a better trade-off than by-productism or why out-group hostility is predictable given theism.

Thirdly, one might question the assumption that if God exists, engaging in religious life will make us better people. Again, we should be careful not to generalize here: the Christian theologian might very well claim that engaging in Christianity makes you a more moral and spiritual person, but this need not apply to all religions. Indeed, given theism the theologian expects some religions to be bad for people—engaging in the worship of ancestors or mystery cults, for instance. Even the effects of Christianity will depend on what one's idea of sanctification is. Some Christian traditions emphasize the possibility of developing virtues through God's grace, whereas others claim that Christians remain sinful all their earthly lives, and do not emphasize individual moral transformation.

CONCLUSION

In the first section, I argued that there are clear differences between cognitive and evolutionary explanations and that the by-product/adaptation divide might be exaggerated. In subsequent sections, I gave a brief overview of debates concerning the religious relevance of CSR theories. My conclusion is rather lukewarm: it seems that the religious relevance of cognitive and evolutionary approaches to religion are (at least in the case of theism) relatively neutral. In other words, when we think about the likelihood of theism as opposed to atheism, CSR gives us very little reason to prefer atheism over theism. On the one hand, the theist has plausible responses to arguments against the truth of theism from CSR. Furthermore, the theist also has some reason to think that CSR has positive religious relevance with respect to certain auxiliary doctrines usually linked with theism. On the other hand, the positive relevance might be cancelled out by the problems that the theist has in responding to some versions of unreliability arguments.

In the final section, I examined the possible religious relevance of EP and CSR together and arrived at a neutral conclusion: even if EP is linked to CSR, its religious relevance for theism collapses into the religious relevance of evolutionary biology. In addition, I examined the religious relevance of the by-product/adaptation divide and concluded that even if we accept the divide in some strong sense (and we should not, I think), the net result is inconclusive. In other words, theism gives us no strong reason to believe that religion is an adaptation instead of a by-product (or vice versa).

The theist might be happy about my lukewarm conclusions. There is a reason, however, why such rejoicing might be premature. What I have examined is whether CSR (and EP) support atheism rather than theism and my conclusion, as we have seen, is fairly neutral: CSR (or EP) all by itself gives us no strong reason to prefer atheism over theism. But it is not just atheism that the theist should be worried about, but naturalism. If we take naturalism to include the claim that everything that exists is natural and can be in principle explained in terms of science and CSR turns out to be true, then this will be positively relevant for naturalism. In other words, naturalistic explanations of religion (if true) might not be direct evidence against theism, but they do give evidence for naturalism, thus making naturalism more plausible than it would be without this evidence.

REFERENCES

Atran, S. 2002. *In Gods We Trust: The Evolutionary Landscape of Religion.* New York: Oxford University Press.

Barkow, J., L. Cosmides, and J. Tooby. Eds. 1992. *The Adapted Mind.* New York: Oxford University Press.

Barrett, J. 2004. *Why Would Anyone Believe in God?* Walnut Creek: AltaMira Press.

——. 2007. Is the spell really broken? Bio-psychological explanations of religion and theistic belief. *Theology and Science* 5: 57–72.

——. 2009. Cognitive science, religion and theology. In *The Believing Primate: Scientific, Philosophical, and Theological Reflections on the Origin of Religion.* Eds. M. Murray and J. Schloss. New York: Oxford University Press, 76–99.

Barrett, J., D. Leech, and A. Visala. 2010 Can religious belief be explained away? Reasons and causes of religious belief. In *The Nature of God—Evolution and Religion Vol. 1.* Ed. U. Frey. Marburg: Tectum Verlag, 75–92.

Boyer, P. 1994. *The Naturalness of Religious Ideas: A Cognitive Theory of Religion.* Berkeley: University of California Press.

——. 2001. *Religion Explained: The Evolutionary Origins of Religious Thought.* New York: Basic Books.

Bulbulia, J. 2009. Religiosity as a mental time travel: Cognitive adaptations for religious behavior. In *The Believing Primate: Scientific, Philosophical, and Theological Reflections on the Origin of Religion.* Eds. M. Murray and J. Schloss. New York: Oxford University Press, 44–75.

Buller, D. 2005. *Adapting Minds: Evolutionary Psychology and the Persistent Quest for Human Nature.* Cambridge, MA: The MIT Press.

Buss, D. 1999. *Evolutionary Psychology: The New Science of the Mind.* Boston: Allyn and Bacon.

Clark, K. J. and J. Barrett. 2010. Reformed epistemology and the cognitive science of religion. *Faith and Philosophy 27*: 174–89.

Clark, K. J. and D. Rabinowitz. 2011. Knowledge and the objection to religious belief from cognitive science. *European Journal for Philosophy of Religion 3*(1): 67–82.

Cohen, E. 2007. *The Mind Possessed: The Cognition of Spirit Possession in Afro-Brasilian Religious Tradition.* Oxford: Oxford University Press.

Cummins, R. 2000. 'How does it work?' versus 'What are the laws?': Two conceptions of psychological explanation. In *Explanation and Cognition.* Eds. F. Keil and R. Wilson, 117–44. Cambridge, MA: The MIT Press.

Day, M. 2007. Let's be realistic: Evolutionary complexity, epistemic probabilism, and the cognitive science of religion. *Harvard Theological Review 100*: 47–64.

Gregersen, N. H. 2006. What theology might learn (and not learn) from evolutionary psychology: A postliberal theologian in conversation with Pascal Boyer. In *Evolution of Rationality: Interdisciplinary Essays in Honor of J. Wentzel Van Huyssteen*. Ed. F. L. Shults. Grand Rapids: William B. Eerdmans, 306–25.

Haines, V. 2007. Evolutionary explanations. In *Philosophy of Anthropology and Sociology*. Eds. S. Turner and M. Risjord. Amsterdam: Elsevier, 246–310.

Johnson, D. and J. Bering. 2009. Hand of God, mind of man: Punishment and cognition in the evolution of cooperation. In *The Believing Primate: Scientific, Philosophical, and Theological Reflections on the Origin of Religion*. Eds. M. Murray and J. Schloss. New York: Oxford University Press, 26–43.

Kahane, G. 2011. Evolutionary debunking arguments. *Nous 45*: 103–25.

Laidlaw, J. 2007. Well-disposed social anthropologist's problem with the 'cognitive science of religion'. In *Religion, Anthropology and Cognitive Science*. Eds. H. Whitehouse and J. Laidlaw. Durham: Carolina Academic Press, 211–46.

Leech, D. and A. Visala. 2011. The cognitive science of religion: A modified theist response. *Religious Studies 47*: 301–16.

——. 2012. How the cognitive science of religion might be relevant for philosophy of religion? In *Scientific Approaches to Philosophy of Religion*. Ed. Y. Nagasawa. London: Palgrave, 165–83.

Mackie, J. L. 1982. *The Miracle of Theism: Arguments For and Against the Existence of God*. Oxford: Clarendon Press.

Mithen, S. 2009. The prehistory of the religious mind. In *Theology, Evolution and the Mind*. Ed. N. Spurway. Newcastle: CSP, 10–30.

Murray, M. 2007. Four arguments that the cognitive psychology of religion undermines the justification of religious belief. In *The Evolution of Religion: Studies, Theories, and Critiques*. Eds. J. Bulbulia et al. Santa Margarita: The Collins Foundation Press, 365–70.

——. 2009. Scientific explanations of religion and the justification of religious belief. In *The Believing Primate: Scientific, Philosophical, and Theological Reflections on the Origin of Religion*. Eds. M. Murray and J. Schloss. New York: Oxford University Press, 168–78.

Murray, M. and A. Goldberg. 2009. Evolutionary accounts of religion: Explaining and explaining away. In *The Believing Primate: Scientific, Philosophical, and Theological Reflections on the Origin of Religion*. Eds. M. Murray and J. Schloss New York: Oxford University Press, 179–99.

Pinker, S. 1997. *How the Mind Works*. New York: Norton.

Plantinga, A. 2000. *Warranted Christian Belief*. New York: Oxford University Press.

Schloss, J. and M. Murray. Eds. 2009. *The Believing Primate: Scientific, Philosophical, and Theological Reflections on the Origin of Religion*. New York: Oxford University Press.

Sosis, R. and C. Alcorta. 2003. Signalling, solidarity, and the sacred: The evolution of religious behaviour. *Evolutionary Anthropology 12*: 264–74.

Van Inwagen, P. 2009. Explaining belief in the supernatural: some thoughts on Paul Bloom's 'Religious Belief as an Evolutionary Accident'. In *The Believing*

Primate: Scientific, Philosophical, and Theological Reflections on the Origin of Religion. Eds. M. Murray and J. Schloss. New York: Oxford University Press, 128–38.
Van Slyke, J. 2011. *The Cognitive Science of Religion*. Farnham: Ashgate.
Visala, A. 2011. *Theism, Naturalism and the Cognitive Study of Religion: Religion Explained?* Aldershot: Ashgate.

5

Extending Evolutionary Accounts of Religion beyond the Mind: Religions as Adaptive Systems

BENJAMIN GRANT PURZYCKI,
OMAR SULTAN HAQUE, AND
RICHARD SOSIS

INTRODUCTION

Scholars and lay people alike have characterized religion as serving a variety of functions. Many have argued that because we are profoundly ignorant of our surroundings, mythologies and religious beliefs developed to satisfy our curiosity by providing explanations for a complex and mysterious universe (e.g. Darwin [1879] 2004, p. 117; Durkheim [1915] 2001, pp. 29–30; Geertz 1973, p. 108). Some propose that religion's essential function is to allow us to cope with death and other forms of psychological suffering (e.g., Becker 1973, pp. 203–4; Spiro 1987, p. 172), while others maintain that religion establishes and sustains the social order (e.g. Radcliffe-Brown 1965).

Within anthropology and related fields, such functionalist explanations have been in decline for decades. Among other deficiencies, functionalist theories often incorrectly assume that particular institutions are essential parts of the societies in which they operate (Collier et al. [1982] 1997, p. 73). Moreover, functionalist theories typically lack any consideration of feedback mechanisms that can account for the maintenance of benefits to social actors (Elster 1983, 2007; Sperber 1996). However, anthropologists, psychologists, and biologists who employ evolutionary models avoid such pitfalls by beginning with and providing explanatory frameworks grounded in the selective retention of useful effects (e.g., Dennett 1995; Owens and Wagner 1992; Smith and Winterhalder 1992). Boyer (2003) nevertheless correctly cautions those seeking evolutionary explanations of religion:

> Most attempts at an evolutionary account of religion have proved unsatisfactory because a single characteristic identified as crucial to the origin of religion is not in fact general. The attempt to find the single evolutionary track for religion is another manifestation of a general urge to identify the single mechanism that motivates religious thought or makes it plausible to believers (p. 123).

We heed Boyer's cautionary advice and approach religion from a more holistic and multivariate perspective than currently offered by the evolutionary and cognitive sciences of religion. Indeed, we argue that an evolutionary approach to religion requires that religion be viewed as a dynamic *system*; isolating particular features of religion for analysis without understanding their influence on the full religious system can be misleading and generate trivial results. The present chapter examines the functional logic of religious systems and explains how they solve adaptive problems. As argued elsewhere (Alcorta and Sosis 2005; Purzycki and Sosis 2009, 2010, 2011; Sosis 2009), the religious system is comprised of cognitive and emotional mechanisms that produce, retain, and motivate commitments to shared supernatural ideas, and behavioural procedures—rituals—that enact, reproduce, and encode these ideas. These components appear to lock into place and form a universal social system with remarkable cross-cultural similarities, as well as predictable differences. Here, we further elaborate how such a system evolves and how the content of its constituent parts adapts to socioecological changes in order to maintain a suitable social niche in which ritual performance can publicly demonstrate group commitments.

BY-PRODUCT ACCOUNTS

There has been considerable debate regarding whether or not religion meets biological criteria for an adaptation (e.g. Bulbulia et al. 2008; Pyysiäinen and Hauser 2010; Sosis 2009). Remarkably, those on all sides of the debate are largely in agreement on the nature of religious cognition and its prosocial effects. Scholars involved in the debate primarily disagree on whether religion is an adaptation, a functional by-product of other adaptations known as an exaptation, or a functionless by-product. The *primary task* of any selectionist analysis is explaining how a particular trait increases the likelihood that its bearer's genes replicate. Adaptations—traits forged by natural selection—often have multiple effects and may be composed of a multitude of other traits that are either a direct or indirect (i.e. a by-product) result of natural selection (e.g. the colour of blood). Moreover, adaptations to past environments can be maladapative—meaning they confer a fitness *disadvantage*—in present ones (e.g. a taste for fatty foods and the overpowering urge to privilege learning over procreation; Richerson and Boyd 2005, pp. 148–90). Alternatively, presently adaptive traits may have been less so in past environments.

Various authors have argued that religious concepts are by-products of evolved cognitive architecture and that religion is therefore an evolutionary by-product (Atran 2002; Barrett 2004; Boyer 1994, 2001; Dawkins 2006; Dennett 2006). These scholars argue that religious concepts co-opt psychological mechanisms that evolved for purposes unrelated to religion, and these concepts tap into emotional and other strategic systems to powerfully motivate behaviour. Boyer (2003), for instance, argues that because religious concepts exist by virtue of 'neural systems' that function for very specific purposes outside of religion, religious thought is a by-product of these systems. Atran (2002) acknowledges that religious behaviour may result in fitness benefits for participants, but like Boyer, he asserts that 'religions are not adaptations and they have no evolutionary functions as such. There is no such entity as 'religion' and not much sense in asking how 'it' evolved. Unlike the case for language, for religion there is no integrated set of cognitive principles that could represent a task-specific evolutionary design' (pp. 264–5). In other words, there is no domain-specific module (or set of genes) for religion and therefore the question of whether or not it is an adaptation is meaningless (Kirkpatrick 1999). The common thread woven throughout by-product accounts of religion is the claim that religion exploits cognitive faculties designed for other, mundane operations, and therefore religion is a by-product of our evolved minds.

By-product accounts that employ this reasoning are prevalent in the evolutionary and cognitive science of religion literature. Among some of the more influential ones are: (a) supernatural agent concepts emerge from mind-detection mechanisms (Barrett 2004; Guthrie 1980, 1995), (b) religious concepts predominantly consist of ideas which violate default assumptions about objects, artifacts, plants, animals, and persons in our world (Atran 2002; Barrett 2004, 2008; Boyer 1994, 2001; Boyer and Ramble 2001; Norenzayan et al. 2006; Pyysiäinen 2004), (c) supernatural agent concepts exploit emotional attachment systems (Granqvist and Kirkpatrick 2008; Granqvist et al. 2010; Kirkpatrick 2005), and (d) ritual behaviour is a by-product of hazard precaution systems (Boyer and Liénard 2006, 2008; Liénard and Boyer 2006). The general logic of these accounts is that because features of religion are analogous to or exploit extant psychological systems, religion is not an evolved trait. While such views offer compelling mechanistic explanations of certain features of religion, they cannot explain four key aspects often encountered in religious systems. First, they fail to account for the recurrent cross-cultural coalescence of many of religion's core features.

Second, they specifically fail to account for coupling of supernatural agent concepts with rituals that are essential to all religious traditions. Third, they fail to account for patterned cross-cultural differences in religious traditions. Fourth, by-product accounts derived from cognitive studies, which focus merely on the mental representations of religious concepts and ritual, operate in an analytical vacuum at the expense of accounting for what religion actually does.

ADAPTATIONIST ACCOUNTS

Humans are composed of many systems, which are comprised of even smaller subsystems. While our stomachs are evolved adaptations for food-processing, the stomach itself is composed of numerous systems of cells, their interactions, and the emergent properties of their collectivity we call 'tissue'. Not only do we consist of various interacting systems, but we also regularly participate in systems far larger than ourselves. No one who embraces evolution by natural selection would argue that the immune system is *not* an adaptation, yet we find no one requiring a particular locus of this adaptation since the system is composed of a number of interacting units and they must function together to maintain the defensive capabilities of organisms against infectious microorganisms. Moreover, no one would deny that our own immune systems are any less adaptive because they served different functions or had different forms in our evolutionary past.

Similarly, what we have been calling the 'religious system' is composed of constituent parts. These parts consist of evolved cognitive, emotional, developmental, and behavioural traits. At its most basic level, this system is composed of ritual behaviours and the supernatural agents to which people appeal. Yet, this system of course has other constituent parts, such as myths, symbols, taboos, sacred values, all of which must operate together in order for individuals to reap the benefits from participation. Adaptive accounts of religion do not disagree with the central and often empirically supported claim of by-product accounts that religious concepts are made possible by evolved psychological mechanisms which evolved as a consequence of selective pressures that are in all probability unrelated to religion. However, an adaptive approach goes a few steps further. First, it gives equal attention to human universals and cross-cultural variation as both are explicable by evolutionary analysis. Second, it emphasizes the functional *effects* of coupling religious concepts, emotions, and behaviours over time on human social interactions. Third, it emphasizes the remarkable recurrence, convergence, and non-random distributions of these features around the world.

RELIGIOUS RITUAL AS A SIGNAL OF COMMITMENT

Pinker and Bloom (1992) suggest that 'Supplementing the criterion of complex design, one can determine whether putatively adaptive structures are correlated with the ecological conditions that make them useful, and under certain circumstances one can actually measure the reproductive success of individuals possessing them to various degrees' (p. 457). Indeed, there is a significant amount of evidence that affiliation with a religious community is positively correlated with fertility (Blume 2009, 2010; Frejka and Westoff 2008; Kaufmann 2010). The mechanisms driving this effect, however, are unclear. The norms and practices of religious communities may serve as solutions to coordination problems for persons who prefer a monogamous, high-fertility mating strategy. Indeed, attitudes towards sex and mating behaviours are very strong predictors of attendance at a house of worship (Weeden et al. 2008). Religious communities may provide places for individuals to find like-minded individuals, such as others with similar mating strategies. This approach, however, fails to explain how religious non-kin and non-mates are also remarkably cooperative with each other. Consideration of religion as a form of communication helps provide a more fruitful approach.

There are undoubtedly forms of communication that we share with our non-human relatives. Animals regularly communicate with others using various signals (Hauser 1996; Searcy and Nowicki 2005). Otte (1974) defines signals as 'behavioral, physiological, or morphological characteristics fashioned or maintained by natural selection because they convey information to other organisms' (p. 385). These signals can vary in form, content, cost, intensity, frequency, and reliability. Human facial expression of emotion, for instance, is a reliable, difficult-to-fake signal of an internal emotional state (Frank 1988; Pinker 1997), but of course there are individual differences in the intensity of expression of emotions. Organisms regularly engage in behaviours that reliably convey various messages.

For instance, Zahavi and Zahavi (1997) detail risky behaviours organisms perform in order to reliably demonstrate their fitness. The so-called 'Handicap Principle' predicts variation in the ways animals will signal their fitness to potential mates, same-sex competitors, and/or predators. These signals are generally reliable because they come at a cost to individuals bearing them and thus accurately communicate the message they are designed to convey. Humans, for instance, regularly engage in conspicuous consumption in order to 'sufficiently put [their] opulence in evidence' (Veblen [1899] 2007, p. 53; see also Miller 2009) just as male bowerbirds collect and arrange attractive colourful trinkets to display hard-to-fake information about their attributes, and so can attract females. While humans can convey their opulence by spending their resources on symbols of wealth (e.g. fast cars, jewels, etc.), bower birds demonstrate their fitness by using energy and time expenditures to amass a hoard of equally apparently useless objects. The real utility, of course, lies in the payoffs for engaging in such behaviour (e.g. prestige and mates, respectively).

Signalling theories of religious behaviour contend that religious rituals also act as signals of an individual's solidarity with the religious tradition and, more importantly, with the specific individuals engaged in that tradition (Alcorta and Sosis 2005; Bulbulia in press; Irons 2001; Sosis and Alcorta 2003). Cooperative relationships face the inherent problem of exploitation by cheaters: those who free-ride (i.e. profit at others' expense) can take advantage of others' generosity (Axelrod 1984; Dawkins and Krebs 1979; Iannaccone 1992, 1995). Religious rituals are one mechanism that minimize free-riding on the cooperative efforts of others. The costs of ritual performance, including time, energy, risk, material, and opportunity costs, enable religious behaviours to serve as reliable signals of in-group commitments. In other words, one who pays the costs of a supernaturally rationalized rite, indicates his or her commitment to the mores of their tradition, and those who share that tradition. Religious rituals can therefore be a means to communicate trustworthiness (Sosis 2005).

Tan and Vogel (2008), for example, found that religiosity predicts trustworthiness and willingness to trust others in economic games. If one is perceived to be more trustworthy, then cooperative relations are likely more reliable, and one can reap the benefit of such cooperative relations. Of course, costly signals are used as signals of commitment in atheistic and secular contexts as well. Tattooing onto one's body, for example, the logo of the atheist 'Brights' movement, one's fraternity letters, Captain Beefheart lyrics, etc., are all demonstrations of costly commitment in secular contexts. However, what costly religious signals do is communicate commitment to social contracts perceived to be mediated by sacred and often eternal beings. There is ever-growing evolutionary modelling (Dow 2008; Henrich 2009; Wildman and Sosis 2011) and empirical evidence that supports the hypothesis that ritual's adaptive function is to strengthen social bonds through costly signals.

Sosis and Bressler (2003; also see Sosis 2000), for example, demonstrated that religious communes outlive secular communes, and that among religious communes, but not secular communes, costly rituals are positively correlated with commune longevity. Religious concepts are therefore likely to be necessary for this prolonged cohesion; when rituals are rationalized in secular terms, they can be explained away. Successful religious concepts are by nature unverifiable and thus provide people with a stable, albeit otherworldly, incentive to participate (Rappaport 1999). As such, religious concepts appear to be a necessary component for the prosocial effects we see in religious groups (Purzycki and Sosis 2010). In related work, Sosis and Ruffle (2003, 2004) found members of religious kibbutzim to be more generous in economic games than their secular counterparts. Similarly, Soler (2012) found that religiosity is positively correlated with in-group generosity among Afro-Brazilians. Moreover, the cost of religious ritual increases with frequency in warfare, suggesting that the greater the organizational need, the greater the ritual cost (Sosis et al. 2007). While costly rituals appear to be important for strengthening cohesion in diverse groups, recent laboratory studies have shown that simple exposure to religious concepts, or concepts that are similarly unverifiable, can promote prosocial behaviour (Norenzayan and Shariff 2008).

SUPERNATURAL PUNISHMENT

As Murray and Moore (2009) note, in many religious contexts there are third-party sanctions against behaviours deemed religiously unacceptable. When violations of sacred taboos may be punished by a community, there is little need for costly rites that overcome the problems of cooperation (Sosis 2005). As discussed below, supernatural agents may serve this function in specific contexts. An evolutionary account of how selection favoured commitments to moralistic gods, known as the 'Supernatural Punishment Hypothesis', predicts that engaging supernatural agent concepts minimizes antisocial behaviour and/or promotes prosocial behaviour (Bering and Johnson 2005; Johnson and Bering 2006; Schloss and Murray 2011).

In a host of important studies, researchers have found that priming of religion-associated concepts increases prosocial behaviour. For example, Shariff and Norenzayan (2007) found that implicit priming of religious concepts positively affected generosity in economic games (though secular authority primes had a similar effect, which suggests a common causal mechanism in reputation monitoring). Moreover, Bering et al. (2005) found that individuals were less likely to cheat in an experiment when primed with a ghost story. Johnson (2005) found a significant correlation between a number of indices of cooperation and commitment to a moralizing high god in a massive cross-cultural sample. While Norenzayan and Shariff (2008) suggest that the cooperative effects of religion make them more attractive to people and this is why individuals commit to them, Bering and Johnson (2005) argue that this evidence suggests an adaptation is at work and that this behaviour may point to something about our species' past.

Over the years, researchers have found again and again that the complexity of state societies correlates with omniscient, moralistic, and often singular deities (Johnson 2005; Lahti 2009; Sanderson 2008; Stark 2001, Swanson 1960; Wallace 1966) but, as noted above, there are many examples of non-state societies with moralistic supernatural agents. The Nuer, for instance, believe in an omniscient moralizing god similar to the Abrahamic deity (Evans-Pritchard 1956); the Ju/'Hoansi of the Kalahari believe their ancestors' spirits know when they misbehave and make them sick if they do (Lee 2003); and the spirits among the Netsilik (Inuit) make people sick for violating taboos (Balikci 1970, p. 226). Nevertheless, the anonymity provided by complex state political systems minimizes accountability. However, if a moralistic omniscient deity is watching, individuals may be less likely to engage in antisocial behaviour. Yet many supernatural agents simply do not care about morality and do not necessarily punish people.

Among the Christian Maisin in Papua New Guinea, for example, 'With the somewhat ambiguous exception of the Christian god, spiritual entities are ultimately amoral. They can aid, harm, or ignore the living as they please' (Barker 2008, p. 122). In Inner Asia, Tyvan spirit-masters are particularly concerned about ritual behaviour and resource preservation and not human morality (Purzycki 2010, 2011). The people of Ifaluk believe in malevolent and benevolent spirits, but divine wrath is explained by virtue of the spirits' malevolence, not necessarily by human misconduct. In fact, human misconduct is explained by spirit possession. On Ifaluk, residents direct individual aggression toward culturally sanctioned targets (Spiro 1952). In such cases, there seems to be little to no supernatural response to human immorality.

One of the underappreciated and therefore underexploited values of this ethnographic research is that it suggests that not only does religious content change to accommodate new socioecological contexts, but that the functions of specific religions may actually shift under new ecological conditions and challenges (Rossano 2006, 2007, 2009; Wright 2009). As noted above, increasing anonymity in densely populated state societies made the monitoring of ritual performance difficult, thus moralistic supernatural agents developed to overcome the new challenges to cooperation. Quite likely, this system was modified from one of stimulating and rationalizing (i.e. explaining costly behaviours with appeals to unverifiable agents) religious ritual—as found in traditional societies—to one of a transcendent monitoring system which has indeed been shown to affect prosocial behaviour (Purzycki and Sosis 2011). However, more specifically, we find considerable variation around the world in the form of supernatural agents (e.g. human-like, animal-like, etc.), the objects of supernatural agents' concerns, the breadth of knowledge, how this variation informs religious behaviour, and how these religious complexes make sense in their socioecological contexts.

RELIGION BEYOND MIND

Bulbulia (2008) characterizes religion as its own 'niche' describing it as 'a system of organized behaviour and knowledge, together with whatever artifacts and other symbolic structures (musical scores, texts, religious architecture) that is supported, retained, improved, and transmitted at least in part because we possess cognitive capacities to believe and morally commit to supernatural realities and purposes' (p. 21). We are all part of interconnected and interpenetrating niche-systems: whether it is the economic, academic, religious, or political, individuals engage in systems which are (at least analytically) isolatable. Our shared representations and behaviours actually forge a context within which we navigate our social lives. Alcorta and Sosis (2005) argue that 'religion may best be understood as an evolved complex of traits incorporating cognitive, affective, behavioural, and developmental elements selected to solve an adaptive problem' (p. 325). These traits make the essential elements of religion possible. As Chomsky (1996) rightly notes, 'Organs do not evolve independently, of course, and a viable organism has to hang together in complicated ways; breeders know how to breed bigger horses, but it won't help if size increases without highly intricate corresponding changes in the brain, the circulatory system, and much more' (p. 16). Like horse breeding, religions can change through the deliberate efforts of religious leaders, but religious traditions also change outside of conscious decisions of individual agents in response to socioecological changes. Communities share religious concepts, and how this sharedness affects behaviour is of utmost significance in understanding religion as a dynamic system.

Perceived sharedness in belief and/or behaviour is arguably necessary to reap the benefits of the religious system. In order to be of any adaptive utility, one's world-view—religious and secular—must have an appropriate degree of compatibility with others' in contexts where (a) such compatibility or lack thereof has fitness consequences, (b) a demonstration of sharedness and commitment is expected, and (c) such demonstrations have costs. As we cannot directly see peers' models and motivations, behaviours often need particular costs in order to demonstrate commitment to such models, regardless of belief. Rituals contain these qualities (Alcorta and Sosis 2005; Sosis 2003, 2006) and their performance can consequently produce adaptive outcomes. But we have little understanding of how shared perceptions are achieved. Nor do adaptive or by-product accounts of religion inform us why the religious system's constituent parts recurrently coalesce across cultures (Sosis 2009). As detailed above, virtually all religions consist of an essential supernatural agent-ritual coupling. Whether the agent is a god, ancestor spirit, totemic animal, demon, etc., one essential component of religion is commitment to a supernatural agent. Supernatural agent concepts are used in contexts where costly rituals need to be rationalized with appeals to unverifiable, highly ambiguous notions. Standard psychological methods employed by cognitively oriented researchers are ill-equipped to address these concerns, but even more dynamic systems approaches are likely to only get so far. There are undoubtedly complex principles at work which current approaches fail to address.

Commenting on standard modelling procedures, Miller and Page (2007) quip that 'The ability to collect and pin to a board all of the insects that live in the garden does little to lend insight into the ecosystem contained therein' (p. 10). Likewise, detailing the essential organs of the religious system will ultimately shed little light on the internal dynamics and co-dependencies of that system and how it interacts with external constraints and influences. Rather, emphasizing the *interactions* between the components of the religious system and the individuals participating in them suggests that the whole is indeed greater than the sum of its parts. There may be emergent properties of these components that are wholly unrecognized by most current approaches to understanding religion (for further discussion, see Holland 1995; Miller and Page 2007, pp. 44–53).

Relatively recent trends in computational modelling, typically employed in artificial intelligence and economics, focus on understanding the laws governing complex adaptive systems. Religion is a complex adaptive system *par excellence* as:

> (*i*) It consists of a network of interacting agents (processes, elements); (*ii*) it exhibits a dynamic, aggregate behavior that emerges from the individual activities of the agents; and (*iii*) its aggregate behavior can be described without a detailed knowledge of the behavior of the individual agents. An agent in such a system is *adaptive* if it satisfies an additional pair of criteria: the actions of the agent in its environment can be assigned a value (performance, payoff, fitness, or the like); and the agent behaves so as to increase this value over time. A complex adaptive system, then, is a complex system containing adaptive agents, networked so that the environment of each adaptive agent includes other agents in the system (Holland and Miller 1991, p. 365).

Even outside of state societies' correlation with moralistic high gods, we find that the *content* of religion changes through time and space in similar and predictable ways in non-state traditions as well; religious traditions show striking similarities given the socioecological context in which they operate (e.g. Sierksma 1963; Snarey 1996; Wallace 1966). Moreover, there is significant evidence to suggest that specific religious traditions function to regulate resource use by virtue of emergent properties which are largely otherwise unknown to individual constituents (e.g. Atran et al. 2002; Lansing 2007; Lansing and Kremer 1993). In other words, properties that exist at the level of the population may indeed affect individuals' fitness. Native scholars have consistently emphasized the inextricable relationship between a people, their religious beliefs, and natural resources (Battiste and Henderson 2000, pp. 97-116; Deloria 2003). In fact, local indigenous populations have globally appealed to their respective religious traditions to resist ecological overexploitation and development (Klubnikin et al. 2000; LaDuke 2005). How do these relationships develop?

Religious systems regularly converge around very practical concerns ranging from life history events (Reynolds and Tanner 1995) and the coordination of access to valuable resources (Atran et al. 2002; Lansing 2007; Lansing and Kremer 1993) to motivating people to organize against a colonial power (Carroll 1975; Wallace 1956) and fostering in-group cooperation (Sosis and Bressler 2003). Religion is—at least in traditional societies—inextricably linked to all social life and may become of heightened significance during time of organizational need (Finkel et al. 2010). Not only does the complex adaptive systems perspective reemphasize the importance of the 'secular utility' of religion (Durkheim [1915] 2001; Wilson 2002), but it also hints at the possibility of a devoted study of the phylogeny of religion based on ecological context that is not reducible to sociological or non-religious categories alone. Through time, religious systems often acclimate in order to solve particular socioecological problems for their constituents. Individual actors in these systems are not necessarily conscious of these changes but may nevertheless reap the benefits from participation. The coming challenges will be to systematically investigate if the specific content of the defining components of religion—commitment to supernatural agents and ritual behaviours—responds to socioecological changes in predictable ways.

CONCLUSION

As many have argued, the evolutionary emphasis on how ecological factors shape response patterns illustrates why the nature–nurture dichotomy is a false one; genotypes do not occur in a vacuum but rather are expressed as phenotypes through ontogenetic processes that are sensitive to environmental conditions (Barkow et al. 1992; Pinker 2003; Ridley 2004). While the search for a 'religious gene' or 'religion module' is likely futile, there is good reason to ask about the heritability of religious psychological propensities. It has been demonstrated in twin-studies, albeit consistently ignored in the social scientific literature, that religiosity (but not religious affiliation) is partly heritable (Bouchard 2004; Bouchard and McGue 2003). As with the significant heritability of political attitudes, the direct as well as indirect action of the genes in question likely gives rise to *predispositions* for religious commitment which the environment mediates. However, these facts about religious commitment may be used in other contexts as well; by-product accounts may demand evidence that these genes are exclusively *for* religious commitment rather than ideological or political commitment generally (see Kirkpatrick 2005). As discussed above, adaptive accounts do not require such exclusivity. Again, while the quest for finding any essential mechanisms that specifically evolved for religion will likely fail, claiming religion to be an exclusively learned phenomenon misrepresents what we now know about human cognition.

The cognitive science of religion raises important new questions about the nature of religious concepts and processing, but its experimental work heavily relies on Western populations. This renders *a priori* generalizations from empirical work difficult (Henrich et al. 2010). The present work begins to map out the space of variation in the expression of religious systems, as this will provide clues about the interdependencies among its elements. This is where a dynamic systems approach will shed more light on all of the components of religion, but also how the content of these components will change in accordance with how people interact with their socioecological environments. Moreover, while there are notable exceptions in the theoretical literature (McCauley and Lawson 2002; Whitehouse 2004), the cognitive sciences of religion have focused primarily on how we represent and process religious concepts and propositions (see Barrett and Keil 1996; Boyer and Ramble 2001). While these approaches shed light on the nature of the religious mind, the conclusion that religion is a by-product does not follow from the idea that religion is possible by virtue of otherwise mundane cognitive mechanisms. Just the fact that religious concepts correlate with ritual and these rituals may have effects on individual fitness indicates that more is at work than by-product accounts suggest.

A dynamic systems approach to religion encompasses both religious cognition and behaviour and treats them as inextricably linked components of an ever-changing social system. The next step in our quest to understand the evolution of the religious system will be to further explore not only how religious systems persist across ecological contexts, but also how they respond to environmental changes. Only by giving equal attention to the nature of the human mind, the natural and social environments, history, and all of the forces which exert influence upon them and their relationships will we be able to come to terms with explaining why humans do the things that they do. More specifically, we need to understand these dynamic relationships in order to understand why humans are religious. Even more specifically, these dynamics inform why *particular* traditions are the way they are, and the natural laws that have forged their paths.

ACKNOWLEDGEMENTS

The authors would like to thank Jessica McCutcheon for providing useful comments on an earlier version of this chapter. Purzycki was supported by the SSHRC-funded Cultural Evolution of Religion Research Consortium (CERC) during the preparation of this chapter. Sosis thanks and acknowledges support from a CTI Fellowship (Evolution and Human Nature) and an ESRC Large Grant (REF RES-060-25-0085) entitled 'Ritual, Community, and Conflict'.

REFERENCES

Alcorta, C. S. and R. Sosis. 2005. Ritual, emotion, and sacred symbols: The evolution of religion as an adaptive complex. *Human Nature 16*(4): 323–59.

Atran, S. 2002. *In Gods We Trust: The Evolutionary Landscape of Religion*. Oxford: Oxford University Press.

Atran, S., D. Medin, N. Ross, E. Lynch, V. Vapnarsky, E. Ucan Ek', J. Coley, C. Timura, and M. Baran. 2002. Folkecology, cultural epidemiology, and the spirit of the commons: A garden experiment in the Maya Lowlands, 1991–2001. *Cultural Anthropology 43*(3): 421–50.

Axelrod, R. 1984. *The Evolution of Cooperation*. New York: Basic Books.

Balikci, A. 1970. *The Netsilik Eskimo*. Prospect Heights: Waveland Press.

Barker, J. 2008. *Ancestral Lines: The Maisin of Papua New Guinea and the Fate of the Rainforest*. Tonawanda: University of Toronto Press.

Barkow, J., L. Cosmides, and J. Tooby. 1992. *The Adapted Mind: Evolutionary Psychology and the Generation of Culture*. New York: Oxford University Press.

Barrett, J. L. 2004. *Why Would Anyone Believe in God?* New York: AltaMira Press.

——. 2008. Coding and quantifying counterintuitiveness in religious concepts: Theoretical and methodological reflections. *Method and Theory in the Study of Religion 20*(4): 308–38.

Barrett, J. L. and F. C. Keil. 1996. Conceptualizing a nonnatural entity: Anthropomorphism in god concepts. *Cognitive Psychology 31*(3): 219–47.

Battiste, M. and J. Y. Henderson. 2000. *Protecting Indigenous Knowledge and Heritage: A Global Challenge*. Saskatoon: Purich Publishing Ltd.

Becker, E. 1973. *The Denial of Death*. New York: Free Press.

Bering, J. M. and D. D. P. Johnson. 2005. 'O Lord... You Perceive My Thoughts from afar': Recursiveness and the evolution of supernatural agency. *Journal of Cognition and Culture 5*: 118–42.

Bering, J. M., K. McLeod, and T. K. Shackelford. 2005. Reasoning about dead agents reveals possible adaptive trends. *Human Nature 16*(4): 360–81.

Blume, M. 2009. The reproductive benefits of religious affiliation. In *The Biological Evolution of Religious Mind and Behavior*. Eds. E. Voland and W. Schiefenhövel. New York: Springer, 117–26.

——. 2010. Von Hayek and the Amish fertility: How religious communities manage to be fruitful and multiply. A case study. In *The Nature of God: Evolution and Religion*. Ed. U. Frey. Marburg: Tectum Verlag, 159–75.

Bouchard, T. J. 2004. Genetic influence on human psychological traits: A survey. *Current Directions in Psychological Science 13*: 148–51.

Bouchard, T. J. and M. McGue. 2003. Genetic and environmental influences on human psychological differences. *Journal of Neurobiology 54*(1): 4–45.

Boyer, P. 1994. *The Naturalness of Religious Ideas: A Cognitive Theory of Religion*. Berkeley: University of California Press.

——. 2001. *Religion Explained: The Evolutionary Origins of Religious Thought*. New York: Basic Books.

——. 2003. Religious thought and behaviour as by-products of brain function. *Trends in Cognitive Sciences 7*(3): 119–24.

Boyer, P. and P. Liénard. 2006. Why ritualized behavior? Precaution systems and action parsing in developmental, pathological, and cultural rituals. *Behavioral and Brain Sciences 29*: 1–56.

——. 2008. Ritual behavior in obsessive and normal individuals. *Current Directions in Psychological Science 17*(4): 291–4.

Boyer, P. and C. Ramble. 2001. Cognitive templates for religious concepts: cross-cultural evidence for recall of counter-intuitive representations. *Cognitive Science 25*(4): 535–64.

Bulbulia, J. 2008. Meme infection or religious niche construction? An adaptationist alternative to the cultural maladaptationist hypothesis. *Method and Theory in the Study of Religion 20*: 1–42.

——. In press. Why 'Costly-signalling' models of religion require cognitive psychology. In *Origins of Religion, Cognition, and Culture*. Eds. A. Geertz and J. Jensen. London: Equinox.

Bulbulia, J., R. Sosis, E. Harris, C. Genet, R. K. Genet, and J. Wyman. Eds. 2008. *The Evolution of Religion: Studies, Theories, and Critiques*. Santa Margarita: Collins Foundation Press.

Carroll, M. P. 1975. Revitalization movements and social structure: Some quantitative tests. *American Sociological Review 40*(3): 389–401.

Chomsky, N. 1996. *Powers and Prospects: Reflections on Human Nature and the Social Order*. Boston: South End Press.

Collier, J. F., M. Z. Rosaldo, and S. Yanagisako. [1982]. 1997. Is there a family? New anthropological views. In *The Gender/Sexuality Reader: Culture, History, Political Economy*. Eds. R. N. Lancaster and M. di Leonardo, 71–81. New York: Routledge.

Darwin, C. [1879] 2004. *The Descent of Man*. New York: Penguin Classics.

Dawkins, R. 2006. *The God Delusion*. New York: Mariner Books.

Dawkins, R. and J. R. Krebs. 1979. Arms races between and within species. *Proceedings of the Royal Society of London. Series B, Biological Sciences 205*(1161): 489–511.

Deloria Jr., V. 2003. *God Is Red: A Native View of Religion*. Golden: Fulcrum Publishing.

Dennett, D. C. 1995. *Darwin's Dangerous Idea: Evolution and the Meanings of Life*. New York: Simon and Schuster.

——. 2006. *Breaking the Spell: Religion as a Natural Phenomenon*. New York: Viking.

Dow, J. 2008. Is religion an evolutionary adaptation? *Journal of Artificial Societies and Social Simulation* 11/2/2. Online, accessed December 9, 2010. Available online at <http://jasss.soc.surrey.ac.uk/11/2/2.html>.

Durkheim, E. [1915] 2001. *The Elementary Forms of Religious Life*. New York: Oxford University Press.

Elster, J. 1983. *Explaining Technical Change: A Case Study in the Philosophy of Science*. Cambridge: Cambridge University Press.

——. 2007. *Explaining Social Behavior: More Nuts and Bolts for the Social Sciences*. Cambridge: Cambridge University Press.

Evans-Pritchard, E. E. 1956. *Nuer Religion*. New York: Oxford University Press.

Finkel, D., P. Swartwout, and R. Sosis. 2010. The Socio-Religious Brain: A Developmental Model. *Proceedings of the British Academy 158*: 287–312.

Frank, R. H. 1988. *Passions within Reason: The Strategic Role of Emotions*. New York: W. W. Norton and Company.

Frejka, T., and C. F. Westoff. 2008. Religion, religiousness and fertility in the US and in Europe. *European Journal of Population 24*: 5–31.

Geertz, C. 1973. *The Interpretation of Cultures*. New York: Basic Books.

Granqvist, P. and L. A. Kirkpatrick. 2008. Attachment and religious representations and behavior. In *Handbook of Attachment: Theory, Research, and Clinical Applications (2nd edition)*. Eds. J. Cassidy and P. R. Shaver. New York: Guilford, 906–33.

Granqvist, P., M. Mikulincer, and P. R. Shaver. 2010. Religion as attachment: Normative processes and individual differences. *Personality and Social Psychology Review 14*: 49–59.

Guthrie, S. E. 1980. A cognitive theory of religion. *Current Anthropology 21*(2): 181–203.

——. 1995. *Faces in the Clouds: A New Theory of Religion*. New York: Oxford University Press.

Hauser, M. D. 1996. *The Evolution of Communication*. Cambridge, MA: The MIT Press.

Henrich, J. 2009. The evolution of costly displays, cooperation, and religion: Credibility enhancing displays and their implications for cultural evolution. *Evolution and Human Behavior 30*: 244–60.

Henrich, J., S. Heine, and A. Norenzayan. 2010. The weirdest people in the world? *Behavioral and Brain Sciences 33*(2–3): 1–75.

Holland, J. H. 1995. *Hidden Order: How Adaptation Builds Complexity*. New York: Basic Books.

Holland, J. H. and J. H. Miller. 1991. Artificial adaptive agents in economic theory. *American Economic Association Papers and Proceedings 81*(2): 365–70.

Iannaccone, L. R. 1992. Sacrifice and stigma: Reducing free-riding in cults, communes, and other collectives. *The Journal of Political Economy 100*(2): 271–91.

——. 1995. Voodoo economics? Reviewing the rational choice approach to religion. *Journal for the Scientific Study of Religion 34*(1): 76–88.

Irons, W. W. 2001. Religion as a hard-to-fake sign of commitment. In *Evolution and the Capacity for Commitment*. Ed. R. Nesse. New York: Russell Sage Foundation, 292–309.

Johnson, D. D. P. 2005. God's punishment and public goods: A test of the supernatural punishment hypothesis in 186 world cultures. *Human Nature 16*: 410–46.

Johnson, D. D. P. and J. Bering. 2006. Hand of God, mind of man: Punishment and cognition in the evolution of cooperation. *Evolutionary Psychology 4*: 219–33.

Kaufmann, E. 2010. *Shall the Religious Inherit the Earth? Demography and Politics in the Twenty-First Century*. London: Profile Books.

Kirkpatrick, L. A. 1999. Toward an evolutionary psychology of religion and personality. *Journal of Personality 67*(6): 921–52.

——. 2005. Evolutionary psychology: An emerging new foundation for the psychology of religion. In *Handbook of the Psychology of Religion and Spirituality*. Eds. R. F. Paloutzian and C. L. Park. New York: Guilford, 101–19.

Klubnikin, K., C. Annett, M. Cherkasova, M. Shishin, and I. Fotieva. 2000. The sacred and the scientific: Traditional ecological knowledge in Siberian river conservation. *Ecological Applications 10*(5): 1296–306.

LaDuke, W. 2005. *Recovering the Sacred: The Power of Naming and Claiming*. Cambridge: South End Press.

Lahti, D. C. 2009. The correlated history of social organization, morality, and religion. In *The Evolution of Religious Mind and Behavior*. Eds. E. Voland and W. Schiefenhövel. New York: Springer, 67–88.

Lansing, J. S. 2007. *Priests and Programmers: Technologies of Power in the Engineered Landscape of Bali*. Princeton: Princeton University Press.

Lansing, J. S. and J. N. Kremer. 1993. Emergent properties of Balinese water temple networks: Coadaptation on a rugged fitness landscape. *American Anthropologist, New Series 95*(1): 97–114.

Lee, Richard B. 2003. *The Dobe Ju/'Hoansi*. Belmont: Wadsworth Publishing.

Liénard, P. and P. Boyer. 2006. Whence collective rituals? A cultural selection model of ritualized behavior. *American Anthropologist 108*(4): 814–27.

McCauley, R. N. and T. Lawson. 2002. *Bringing Ritual to Mind: Psychological Foundations of Cultural Forms*. Cambridge: Cambridge University Press.

Miller, G. 2009. *Spent: Sex, Evolution, and Consumer Behavior*. New York: Viking.

Miller, J. H. and S. E. Page. 2007. *Complex Adaptive Systems: An Introduction to Computational Models of Social Life*. Princeton: Princeton University Press.

Murray, M. J., and L. Moore. 2009. Costly signaling and the origin of religion. *Journal of Cognition and Culture 9*(3): 225–45.

Norenzayan, A. and A. F. Shariff. 2008. The origin and evolution of religious prosociality. *Science 322*: 58–62.

Norenzayan, A., S. Atran, J. Faulkner, and M. Schaller. 2006. Memory and mystery: The cultural selection of minimally counterintuitive narratives. *Cognitive Science 30*: 1–30.

Otte, D. 1974. Effects and functions in the evolution of signaling systems. *Annual Review of Ecology and Systematics 5*(1): 385–417.

Owens, D. A. and M. Wagner. 1992. *Progress in Modern Psychology: The Legacy of American Functionalism*. Westport: Praeger Publishers.

Pinker, S. 2003. *The Blank Slate: The Modern Denial of Human Nature*. New York: Penguin.

——. 1997. *How the Mind Works*. New York: Norton.

Pinker, S. and P. Bloom. 1992. Natural language and natural selection. In *The Adapted Mind: Evolutionary Psychology and the Generation of Culture*. Eds. J. H. Barkow, L. Cosmides, and J. Tooby. New York: Oxford University Press, 451–93.

Purzycki, B. G. 2010. Spirit masters, ritual cairns, and the adaptive religious system in Tyva. *Sibirica 9*(2): 21–47.

——. 2011. Tyvan cher eezi and the socioecological constraints of supernatural agents' minds. *Religion, Brain and Behavior 1*(1): 31.

Purzycki, B. G. and R. Sosis. 2009. The religious system as adaptive: Cognitive flexibility, public displays, and acceptance. In *The Biological Evolution of Religious Mind and Behavior*. Eds. E. Voland and W. Schiefenhövel. New York: Springer, 243–56.

——. 2010. Religious concepts as necessary components of the adaptive religious system. In *The Nature of God: Evolution and Religion*. Ed. U. Frey. Marburg: Tectum Verlag, 37–59.

——. 2011. Our gods: Variation in supernatural minds. In *Essential building blocks of human nature*. Eds. U. Frey, C. Störmer, and K. P. Willführ. New York: Springer-Verlag, 77–93.

Pyysiäinen, I. 2004. *Magic, Miracles, and Religion: A Scientist's Perspective*. Walnut Creek: AltaMira Press.

Pyysiäinen, I. and M. Hauser. 2010. The origins of religion: evolved adaptation or byproduct? *Trends in Cognitive Sciences 14*(3): 104–9.

Radcliffe-Brown, A. R. 1965. *Structure and Function in Primitive Society: Essays and Addresses*. New York: Free Press.

Rappaport, R. A. 1999. *Ritual and Religion in the Making of Humanity*. Cambridge: Cambridge University Press.

Reynolds, V. and R. Tanner. 1995. *The Social Ecology of Religion (2nd edition)*. New York: Oxford University Press.

Richerson, P. J. and R. Boyd. 2005. *Not by Genes Alone: How Culture Transformed Human Evolution*. Chicago: The University of Chicago Press.

Ridley, M. 2004. *The Agile Gene: How Nature Turns on Nurture*. New York: Harper Perennial.

Rossano, M. J. 2006. The religious mind and the evolution of religion. *Review of General Psychology 10*: 346–64.

——. 2007. Supernaturalizing social life: Religion and the evolution of human cooperation. *Human Nature 18*: 272–94.

——. 2009. Ritual behaviour and the origins of modern cognition. *Cambridge Archaeological Journal 19*: 243–56.

Sanderson, S. K. 2008. Religious attachment theory and the biosocial evolution of the major world religions. In *The Evolution of Religion: Studies, Theories, and Critiques*. Eds. J. Bulbulia, R. Sosis, E. Harris, C. Genet, R. K. Genet, and J. Wyman. Collins Foundation Press: Santa Margarita, 67–72.

Schloss, J. P. and M. J. Murray. 2011. Evolutionary accounts of belief in supernatural punishment: a critical review. *Religion, Brain and Behavior 1*(1): 46.

Searcy, W. A. and S. Nowicki. 2005. *The Evolution of Animal Communication: Reliability and Deception in Signaling Systems*. Princeton: Princeton University Press.

Shariff, A. F. and A. Norenzayan. 2007. God is watching you: Priming god concepts increases prosocial behavior in an anonymous economic game. *Psychological Science 18*: 803–9.

Sierksma, F. 1963. Sacred cairns in pastoral cultures. *History of Religions 2*(2): 227–41.

Smith, E. A. and B. Winterhalder. Eds. 1992. *Evolutionary Ecology and Human Behavior*. New York: Aldine de Gruyter.

Snarey, J. 1996. The natural environment's impact upon religious ethics: A cross-cultural study. *Journal for the Scientific Study of Religion 80*: 85–96.

Soler, M. 2012. Costly signaling, ritual and cooperation: Evidence from Candomblé, an Afro-Brazilian religion. *Evolution and Human Behavior 33*(4): 346–356.

Sosis, R. 2000. Religion and intra-group cooperation: preliminary results of a comparative analysis of utopian communities *Cross-Cultural Research 34*: 70–87.

——. 2003. Why aren't we all Hutterites? Costly signaling theory and religious behavior. *Human Nature 14*: 91–127.

——. 2005. Does religion promote trust? The role of signaling, reputation, and punishment. *Interdisciplinary Journal of Research on Religion 1*: 1–30.

——. 2006. Religious behaviors, badges, and bans: signaling theory and the evolution of religion. In *Where God and Science Meet: How Brain and Evolutionary Studies Alter*

Our Understanding of Religion, Volume 1: Evolution, Genes, and the Religious Brain. Ed. P. McNamara. Westport: Praeger, 61–86.

——. 2009. The adaptationist-byproduct debate on the evolution of religion: Five misunderstandings of the adaptationist program. *Journal of Cognition and Culture 9*: 315–32.

Sosis, R. and C. Alcorta. 2003. Signaling, solidarity and the sacred: the evolution of religious behavior. *Evolutionary Anthropology 12*: 264–74.

Sosis, R. and E. Bressler. 2003. Cooperation and commune longevity: a test of the costly signaling theory of religion. *Cross-Cultural Research 37*: 211–39.

Sosis, R. and B. Ruffle. 2003. Religious ritual and cooperation: Testing for a relationship on Israeli religious and secular kibbutzim. *Current Anthropology 44*: 713–22.

——. 2004. Ideology, religion, and the evolution of cooperation: Field tests on Israeli kibbutzim. *Research in Economic Anthropology 23*: 89–117.

Sosis, R., H. Kress, and J. Boster. 2007. Scars for war: Evaluating alternative signaling explanations for cross-cultural variance in Ritual Costs. *Evolution and Human Behavior 28*: 234–47.

Sperber, D. 1996. *Explaining Culture: A Naturalistic Approach.* Malden: Blackwell.

Spiro, M. E. 1952. Ghosts, Ifaluk, and teleological functionalism. *American Anthropologist, New Series 54*(4): 497–503.

——. 1987. Collective representations and mental representations in religious symbol systems, in B. Kilborne and L. L. Langness, (eds), *Culture and Human Nature: Theoretical Papers of Melford E. Spiro.* Chicago: University of Chicago Press, 161–84.

Stark, R. 2001. Gods, rituals, and the moral order. *Journal for the Scientific Study of Religion 40*(4): 619–36.

Swanson, G. E. 1960. *The Birth of the Gods: The Origin of Primitive Beliefs.* Ann Arbor: University of Michigan Press.

Tan, J. H. and C. Vogel. 2008. Religion and trust: An experimental study. *Journal of Economic Psychology 29*(6): 832–48.

Veblen, T. 2007. *The Theory of the Leisure Class.* New York: Oxford University Press.

Wallace, A. F. C. 1956. Revitalization movements. *American Anthropologist 58*(2): 264–81.

——. 1966. *Religion: An Anthropological View.* New York: McGraw-Hill.

Weeden, J., A. B. Cohen, and D. T. Kenrick. 2008. Religious attendance as reproductive support. *Evolution & Human Behavior 29*: 327–34.

Whitehouse, H. 2004. *Modes of Religiosity: A Cognitive Theory of Religious Transmission.* Walnut Creek: AltaMira Press.

Wildman, W. J. and R. Sosis. 2011. Stability of groups with costly beliefs and practices. *Journal of Artificial Societies and Social Simulation 14*(3): 6.

Wilson, D. S. 2002. *Darwin's Cathedral: Evolution, Religion, and the Nature of Society.* Chicago: University Of Chicago Press.

Wright, R. 2009. *The Evolution of God.* New York: Little, Brown and Company.

Zahavi, A. and A. Zahavi. 1997. *The Handicap Principle: a Missing Piece of Darwin's Puzzle.* New York: Oxford University Press.

6

Skilful Engagement and the 'Effort after Value': An Axiological Theory of the Origins of Religion

NATHANIEL F. BARRETT

INTRODUCTION: THE TURN FROM TRUTHFUL BELIEF TO VALUE-RICH ENGAGEMENT

The philosopher Alfred North Whitehead is credited with saying that it is better to be interesting than to be true (1967, p. 244). Surely only a sheltered intellectual would endorse such a frivolous notion! Real life—that is, life outside the ivory tower—presents us with hard facts that we neglect at our peril. It would be interesting to live in trees, for instance, but the force of gravity dictates otherwise. However, Whitehead's point was not to derogate truth but to point out its dependence on *value*. Truths that lack value are hardly worthy of the name: they are superficial generalities and bland platitudes: trivial statements that give us little traction on the aforementioned facts of life. More substantial truths depend on the attainment of value-rich experience, because value is essential to *engagement*.

The connection between value and engagement is easier to grasp once we have distinguished engagement from interpretation. Although our engagement with the world around us is intimately related to our beliefs about the world, they are not the same. Beliefs can be true or false, while engagement is better described as deep or shallow, wide or narrow, rich or poor. If we fail to engage something we are not so much mistaken about it as oblivious. Moreover it is possible to engage something and at the same time be mistaken about its meaning. For instance, in face-to-face conversation we often engage non-verbal cues like body language and tone of voice without thinking about them, and when we do think about them we are sometimes mistaken about what they mean—taking nervousness for hostility, for example. But such mistakes are different from the obliviousness of simply failing to register these cues altogether.

Engagement, then, is at least as important as interpretation. And the quality of engagement—rich or poor, wide or narrow, deep or shallow—matters so that *what* we engage is as important as *how* we engage it. And herein lies the crucial role of value in cognition, because it turns out that the same terms that can be used to describe experiences that are full of value can also be used to describe engagement that is rich, wide, and deep. Indeed, because of its importance for engagement, one can go so far as to say that the experience of value is not something added to cognition but its essence and foundation.

In light of their importance, it makes sense that our habitual ways of engaging with the world cannot be changed at will. But they can be changed, and the methods and techniques by which we gain leverage on them are of vital importance to human life. The methods most celebrated today belong to the sphere of interpretation: tools of reflection, comparison, and criticism that enable us to sift through the beliefs that both guide and stem from habits of engagement. Our powers of interpretation seem to have grown enormously in recent centuries, and their impact on human life cannot be overstated. But we should not neglect other, less intellectual means of shaping engagement: these include all the various disciplines and practices that work directly on the enrichment of experience, especially training in the arts, music, dance, but also any form of 'enskilment' that trains the body to interact with the environment in some special way. The most highly developed of these techniques are powerful and comprehensive enough to shape an entire way of life. Their development in early human history must have been revolutionary, at least as important as any technological breakthrough. The earliest of these 'technologies of the imagination' were forerunners of the various distinct spheres of value-rich experience—including religion—whose flowering marks the emergence of civilization.

Accordingly, in what follows I attempt a radical reframing of the question of religion's evolutionary origins in terms that make value-rich engagement, rather than truthful belief, the primary driving force of human biocultural evolution. The philosophical basis of this approach is the systematic axiological theory of human experience developed by Robert Cummings Neville (1981, 1989, 1995). Neville's turn toward value and engagement has profound consequences for how we think about religion, as evidenced by his own theory of symbolic engagement (1996, 2006). Here my task is to show how it leads to a new perspective on the cognitive, behavioural, and experiential roots of religion. In this task I am aided by the field of 'ecological psychology' (Gibson 1986), perhaps the only existing branch of experimental psychology that has attempted to register the central importance of value and engagement.

My basic proposal is to reconceive the emergence of religious traditions as historically recent developments of a deep biocultural evolutionary process whose most basic orientation is, in the words of the late ecological psychologist Edward S. Reed, the 'effort after value and meaning' (1996, pp. 96–110). This theory has the potential to recast many important episodes of religious history as processes of value discovery. For the purposes of this essay, however, I am concerned only to show how techniques for the cultivation of value-rich experience could have arisen within the seemingly mundane subsistence activities that constitute everyday life in a hunter-gatherer society. These activities demand sensitivity and skill, and my central claim is that experiences of intense value—not religious *per se*, but rich with religious possibilities—are intrinsic to skilful engagement.

Establishing the feasibility of the value-centred approach to human experience and cognition is the main burden of the following argument, and depends on showing that the interactive picture of cognition offered by ecological psychology (e.g. Chemero 2009) is appropriately described in axiological terms as the 'carryover of value' (Neville 1989). Then, to connect this theory of experience with religion, I turn to the work of the anthropologist Tim Ingold, who has drawn on ecological psychology to re-interpret the animistic cultures of hunter-gatherer societies (2000). Anticipating the point that I wish to make about religion, Ingold has argued that these cultures should be appreciated as forms of engagement rather than systems of belief.

It would be a mistake, however, to view such cultures simply as proto-religions. Rather, in light of Ingold's work, what we now distinguish as religion, art, music, and other distinct fields of imaginative discovery appear to be derivative products of what was originally a unitary pursuit of values—a way of life—enabled and enriched by the development of various interrelated 'technologies of the imagination'. Importantly, these early forms of engagement were not separated off from mundane techniques of subsistence and survival: all of life was a seamless endeavour for the kind of value-rich experience that combines skill mastery with the deep satisfaction of 'being at home' as two sides of a coin. The gradual differentiation of religion as a distinct sphere of values over the last several thousand years—leading eventually to the view that religion is the unique and *sui generis* source of value per se—is a crucial phase of its continuing evolution that has largely obscured or even, in some cases, severed its roots in engagement. Thus an evolutionary perspective that prioritizes value and engagement may offer fresh insights into ongoing developments of 'religion in modernity'.

COMPARISON WITH STANDARD COGNITIVE-EVOLUTIONARY THEORIES

It is important to recognize that the turn from belief to value not only goes against the grain of the most common understanding of religion; it also runs counter to mainstream cognitive theory. It thus presents a radical alternative to the kinds of cognitive-evolutionary theories that have until recently dominated evolutionary discussions of religion (cf. Bellah 2011).

The most prominent cognitive-evolutionary theories seek to explain the apparently widespread persistence throughout human history of supernatural beliefs (Atran 2006; Tremlin 2006). The typical strategy of this approach is to show how this seemingly universal cognitive liability can be explained in terms of (1) inferential mechanisms distinctive to the human brain and (2) selection pressures in our ancestral environment under which these mechanisms evolved. It is important to recognize that this research is driven by a view of religious belief as highly problematic: an evolutionary theory of the propensity for supernatural belief is called for precisely because, from an evolutionary standpoint, it stands out as curiously maladaptive.

The evolutionary problem of religious belief stems from the view of cognition as inferences made from a limited input of sense data, i.e. 'information processing'. From this view, the evolution of human cognition is pictured as the development of a universal set of inferential mechanisms, which can then be used to explain species-wide propensities for certain beliefs. Given the rather precarious existence of proto-humans, it seems reasonable to assume that selection pressures on the ability to acquire reliably accurate beliefs about the environment were quite strong. And so the seemingly universal susceptibility to counterfactual beliefs—evidenced as much by 'primitive' animism as by the most sophisticated theologies—is highly problematic from an evolutionary perspective. How could nature let us get away with such foolishness?

Most cognitive-evolutionary theories answer this question by claiming that the propensity for supernatural belief was in fact selected for, directly or indirectly, either because it was adaptive in ancestral environments or because it emerged as a by-product of inferential mechanisms that were adaptive. For example, Scott Atran has proposed that our promiscuous ascriptions of agency are the by-product of a 'hair-triggered' agency detector that evolved under pressure to sense and react quickly to hidden predators (Atran 2006).

I have criticized this approach elsewhere at length (2010). Here I would like to focus instead on how the turn to value reverses the priorities of evolutionary explanation. For the standard cognitive-evolutionary approach, the experience of value is not really a part of cognition but something added to it, a motivational overlay. You might say that value is the icing on the cake, which is the generation of belief. In contrast, for the axiological approach that I am proposing, the priority is reversed: experience of value is the cake, and belief is the icing.

Accordingly, for example, animistic hunter-gatherer cultures are not, primarily, distinctive systems of religious or proto-religious belief, but distinctive forms of engagement with selected values—in other words, particular ways of life within particular environments. It is of course possible to extract from such ways of life a system of beliefs, but such systems are highly misleading. The hunter-gatherer who engages animals as personal agents is not *believing* something about those animals. This is not because he has yet to enter the theoretical or self-critical stage of thought, but because successful forms of engagement are not so much interpretations of experience as inroads to experience. That is to say, forms of engagement do not create value so much as open up new venues for its discovery.

Sentences like the previous one suggest a return to a pre-modern mindset for which cherished values are assumed to have absolute and universal status. But in fact the axiological view of cognition provides a different way of understanding the cultural and historical relativity of values, including what it means to suffer a crisis of values. What distinguishes the modern crisis of values, I suggest, is the way that it is usually interpreted self-consciously as a crisis of belief, as if value can only be *ascribed* to things with more or less confidence. The possibility that differences of value might be the result of different kinds of engagement is seldom even considered.

Nowhere is this characteristically modern limitation more clearly evident than in discussions of religion. The development of modern thought about religion—including the concept of religion itself—has largely taken for granted a dichotomy of facts and values, such that it is now commonplace to think of religion as a way of supplying value and meaning to an otherwise valueless and meaningless existence. Moreover it is commonly assumed that the religious supply of value and meaning is dependent on the prior attainment of religious belief. Note that this way of thinking about religious value presumes that things actually encountered in the world cannot be sources of value, even if that is how they are experienced. This 'constructivist' interpretation of value is so deeply ingrained within so many modern cultures, scientific and humanist, intellectual and popular, that it is now almost impossible to question, the ubiquitous presence of value in experience (and the protests of a few philosophers) notwithstanding. As long as science is conceived as dealing only with a world of value-free facts, the notion that value is ascribed rather than engaged is practically unassailable.

RECONCEIVING INTERACTIVE COGNITION AS ENGAGEMENT WITH VALUE

At long last, however, recent developments within cognitive science show signs of bringing this fundamental assumption of modernity into question. As expressed by Rodney Brooks' catchphrase, 'the world is its own best representation', one of the most prominent themes of the 'embodied' movement in cognitive science is its insistence that the mind relies wherever possible on readily available information in the environment. But information is not the same as value, and so far the extension of mind into the world has not yet been applied to our experience of value, even by the more 'radical' kinds of embodied cognitive science known as dynamical systems theory (Kelso 1995) and ecological psychology (Chemero 2009). But by their attention to value and meaning (Reed 1996) these distinctive approaches have at least made it possible to pose the question: given that value is both prominent and pervasive in experience, what is available in the environment to be 'picked up' as value?

This seems like an ill-posed question as long as we regard value only in terms of the priorities and interests that we bring to our engagement with the world. In this section, however, I will argue that skilful engagement is partly dependent on dynamic properties of the environment that are appropriately described as values in their own right.

Is this an argument for the objective status of value? Not in any simple sense. The central claim is that value can be understood in dynamic structural terms that apply both to phenomenological traits of our experience of value and to objectively verifiable features of the environment that are engaged by this experience. No claim is made, however, about the *formal correspondence* of these value structures. Rather value structures of experience are *specified* by value structures of the environment through an extended process of interaction. Nevertheless, especially when this process of specification is manifested as skilled interaction, it is possible to say that values of the environment are 'carried over' into experience. The precise respects in which values of the environment are carried over, and the question of whether this carry-over serves appropriately to guide conduct, is the business of interpretation (Neville 1989). Here I am only interested in explaining what it means to understand value-rich experience as skilful engagement with a value-rich environment.

Obviously much of this argument turns on the 'structure' of value. Does value really have structure? Some ways of construing value reduce it to qualitative properties, but when taken separately these are mere abstractions. Value in experience always combines qualities, and the manner of combination always has some kind of structure. The particular qualities that combine to make up a value are of course important to that value's identity, but here I wish to focus only on structural properties common to all values. Two things are important to keep in mind: first, the structure common to all experiences of value is not formal (in fact value structure is more general than form); second, although at times it is helpful to 'freeze' our experience of value and analyze it as if it were a snapshot, the value structure of experience is better thought of in dynamic terms, as something that unfolds and develops in time.

With these caveats in mind, let us consider value structure in terms of complexity, integrity, contrast, and hierarchy. These features are not equally prominent in every experience of value—in some experiences they might be quite minimal—but every value can be analyzed in their terms. Complexity refers to the fact that all values somehow combine diverse elements, as just described. Integrity is the other side of this very same feature: the diverse elements that make up a value must be brought together somehow. Contrast refers to the relational meaningfulness achieved by diverse elements together in some form of arrangement. Contrast is perhaps the best indication of what value essentially is, both as an achievement of intrinsic worth and as something added to sheer diversity. Every value can be analyzed as both giving value to and receiving value from its components: it is a way of having diverse things together that uniquely defines and exhibits the diverse values of these things as a components of a complex contrast. Finally, as just implied, every value is a *valuing*: the achievement of contrast requires the imposition of hierarchy so that some componential differences participate in intense, focused contrasts while others form a vague and trivial background. Again, the hierarchy of value structure is not necessarily a formal property. It is better conceived in terms of importance, or better yet, in terms of effective participation.

All experience displays these structural features, but in experiences of intrinsic value, or experiences that we would describe as value-rich, they are especially distinct. There are at least two general types of value-rich experience: one is the kind that regards a particular object (often a person) as intrinsically valuable, the other is experience that is itself enjoyed as intrinsically valuable. The psychologist Mihaly Csikszentmihalyi has famously dubbed the latter type 'flow' (1990). Here I wish to emphasize features common to all such experiences: flow can be described simply as an intensification of the value structure inherent to all experience, and the appreciation of a particular object as intrinsically valuable should be understood dynamically as an extended product of interaction, just as John Dewey claimed in his classic, *Art as Experience* (1934).

These are phenomenological claims that can be further clarified, and for which abundant examples can be given, but ultimately they must appeal to experience. Either these features can be appreciated in experience or the argument is dead in the water. To make matters worse, here the role of these features in aesthetic appreciation and their relationship to other accounts of value and beauty can only be pointed out in passing (e.g. Turner 1992). For the sake of the argument, let the aesthetic importance of these features be taken for granted. The matter at hand is to show that what we might normally regard 'merely' as features of aesthetic enjoyment also play an important role in ongoing, interactive, and skilful engagement with a complex environment.

Let us begin with a closer look at engagement. Importantly, engagement is different from mere causal influence *per se*: we do not engage, for example, with the molecules that we ingest in our food (Reed 1996, p. 17). Rather, engagement is the selective coupling of an organism's intrinsic dynamics as a whole—its behavioural dynamics—with features of the environment as these are encountered by the organism. As a complex organism explores its environment, it engages through its sensory organs with continuous flows of ambient energy, and it does so not in piecemeal fashion, but by coupling these flows of stimulation with its ongoing self-constitution as an organism. The causal relationship entailed by engagement is thus irreducibly interactive: while features of the environment can be said to determine the intrinsic dynamics of the organism, the organism not only selects the features that it engages, it maintains the causal interface of engagement through its active explorations of the environment. The interactive quality of engagement is analogous to sailing: while the wind propels the boat, the sailor decides how to use the wind to progress along her intended course, continually repositioning the boat and trimming its sails.

'Coupling' is the standard term for describing a reciprocal causal relationship between two dynamic systems, but I prefer 'engagement' because it better conveys the massive complexity and constant fluidity of the special coupling between organism and environment that constitutes what we call 'experience'. However, though special in many respects, experiential engagement is not essentially different from other kinds of coupling: it brings highly complex features of the environment into causal interaction with a subtly differentiated repertoire of behaviour.

Now let us consider that, to be effective, experiential engagement must exhibit the traits of value structure: complexity, integrity, contrast, and hierarchy. Why is this? At first glance it might seem that the limitations imposed by the features of value structure would present a liability: if not for these restrictions, perhaps the organism could take in even more detail. But while the organism needs to take in as much information as it can, it also needs to be able to act decisively on this information. And for the optimal combination of detail and decisiveness the achievement of complex but focused contrasts is essential. In fact the contrasts of value-rich experiential engagement are not presented *to* the organism at all; rather they *are* the organism as constituted by its trajectory through an evolving state space—its path through experience.

The link between value and skilful engagement is brought one step closer if we shift our attention to valuation as a dynamic process. For the ongoing determination of value structure that constitutes a path through experience, the hierarchy of value is especially important. Not only is it a condition of focused contrasts, which allow the organism to respond in a unitary and decisive way to complex situations, it also gives to experience the character of *depth*. In the continuous unfolding of experience, the character of depth refers to the imminent determinability of 'background' features at the margins of experience: the potential of vague and trivial features to become important participants in newly focused contrasts. While experience always has the hierarchical structure of relatively clear and distinct features standing out within a more vaguely determined field (even relatively unfocused, so-called 'lantern consciousness' must display this structure to some degree), the content of this structure is in constant flux. Most importantly, from the point of view of the experiencing subject, the hierarchical value structure of experience is something *to be explored*. As a basic phenomenological trait of experience, value is not enjoyed passively but *entered into*.

Thus value structure gives to experience what is perhaps its most important feature: its eminent and endless 'explorability'. This aspect of human experience is the subjective side of what can be described objectively as its amazing evolvability, evidenced by the fact that both conscious feelings and neural dynamics are never exactly the same twice (Edelman and Tononi 2000). Indeed this exploratory dimension is not a luxury added to experience but its very essence. Moreover, exploration is not a preliminary phase of skilful engagement; rather skilful engagement is distinguished by the extent to which it opens up and sustains exploration. A master musician, for instance, never ceases to explore new colours and textures of sound. Well-cultivated habits of engagement, therefore, constitute much more than rich couplings: they constitute open-ended pathways of evolving interaction.

Still, even if the value structure of experience is so important for engagement, why should it depend on a value-rich environment? Is it not an independent achievement of the organism, dictated by its own personal needs and interests?

Most views of experience presume that its distinctive structures are constructed because the sense data that is given to experience is largely unstructured. The constructive process of imaginative synthesis, posited by Kant as a condition for the very possibility of experience, is widely taken for granted today. It is reinforced within cognitive science by the picture of sensation as the bombardment of sensory surfaces by particulate stimuli. For this view, regardless of how the world is structured, as actually encountered in sensation it has been pulverized into a welter of sense data and must be reconstructed by elaborate inferential processes of a many-tiered cognitive architecture. From this perspective it seems possible to grant all of the points just made about the value structure of experience and its importance for engagement and yet still hold that value is given to the world, not discovered within it.

The first hurdle to overcome, then, is the assumption that the world as encountered at the level of sensory surfaces is necessarily broken down into unstructured data-points. If, instead, we assume that what we encounter at this most basic level are continuous *flows* of stimulation, then the question becomes: *what structure is directly available through these flows?* The answer, it turns out, is quite a lot. For visual perception, environments are full of variously shaped and textured surfaces that specify complex patterns of ambient light (Gibson 1986), which, in turn, when explored by an appropriately attuned organism, specify intricately structured flows of optical stimulation. These flows contain loads of meaningfully structured information that is unavailable in discrete data.

For instance, one of the most studied of these ecological phenomena—so called because they are generated by interaction between organism and environment—is the optical flow that radiates outward from the centre of the field of vision as an animal approaches a surface. This common but seemingly incidental phenomenon specifies a very important piece of information—time to impact—and it has been demonstrated to be of critical importance to the coordination of animal behaviour, including, for instance, the gannet's plummeting dive into the ocean (Lee and Reddish 1981). What is important to realize about such examples is that fine tuning of behaviour—in the case of the gannet, the last-second retraction of its wings before impact with the ocean surface—need not be achieved through calculations performed on sense data but can be directly geared, so to speak, to optical flow. Note the elegance, efficiency, and, no doubt, the evolutionary advantage of directly engaging information that is important to the successful coordination of action. After all, what is of interest to a predator is not the speed or distance of its prey, but more complex properties of the evolving situation that includes both predator and prey—properties like time to impact.

In fact a fascinating variety of experiments indicates that perception in general is geared toward a wide range of complex, dynamic, and relational properties that are meaningful for action (Chemero 2009). What we are most interested to perceive are not the simple, physical properties of things (mass, etc.) but the kinds of actions they afford in relation to us: that is, their 'affordances'. For example, when rotating a large object in our hand we are not directly gauging size or weight but a more distributed, action-oriented property that could be described as 'wieldiness' (Chemero 2009, pp. 156-8).

Of course, just because some experiments seem to show that complex, distributed properties can be perceived directly does not prove that they are. The purpose of this argument, however, is not to force cognitive constructivists to relinquish their position but to show the plausibility of an alternative. And my most ambitious point, that value-rich experience is dependent on engagement with a value-rich environment, is still before us. Ecological psychology has flirted with this position but never embraced it. The writings of James Gibson combined direct realism with a thoroughly relational view of the properties engaged through perception, and the ambiguous subjective/objective 'mutualism' of such properties has divided interpreters of his ecological approach (Costall 1999; Reed 1996).

At the very least, within a Gibsonian framework it can be confidently maintained that the meaning and value of affordances are *unconstructed*. Ecological meaning and value simply *obtain* as possibilities of interaction between an organism and certain features of its environment (cf. Chemero 2009). This alone may constitute an important advantage over the mainstream constructivist position. It does not, however, do justice to the phenomenological character of value as something discovered, and it falls short as a theory of engagement.

This is because it lacks the exploratory and evolutionary character that I have claimed is an essential trait of experiential engagement. If ecological value and meaning simply obtain for the perceiver, they are not *entered into*, and this quality of depth is an essential phenomenological and functional character of experience (Barrett 2009).

On the other hand, to claim that value-rich engagement is dependent on a value-rich environment seems to go beyond the ecological enclosure of relational properties defined within a context of interaction, straining toward some kind of metaphysical 'thing-in-itself'. What case can be made for the independent achievement of value that is grounded in interaction?

Here I can only offer a couple of suggestions. The first has already been implied by my insistence on the connection between the value structure of experience and its exploratory character. As a dynamic property, this value structure is not something that simply obtains; rather it evolves, and its evolution, while partly determined by intrinsic dynamics of the organism, is geared to flows of stimulation. Therefore, I propose that stimulation flows that are capable of specifying the complex dynamics of value-rich experience must be of equivalent complexity: they need not have the *same* value structure, but they must have value structure of some kind. That is to say, if we view the causal conditions of the phenomenological character of value structure in terms of selective *coupling* between organism and environment (rather than constructive synthesis), then value-rich experience is dependent on the coupling of systems of equivalent complexity.

My second suggestion shifts from the causal conditions of experience to the cognitive benefits of engaging properties of the environment that have value structure. Value structures, I propose, are 'handles' on a complex and changing world: they are exactly the sorts of features that an organism should seek to engage for the purpose of *skilful* interaction. Looking back to the general features of value structure, we find that they can also apply to features that are important within their own causal context: hierarchy, for instance, can refer to the way in which the dynamic trajectory of a complex system is governed by relatively few collective variables. Dynamical systems theory and complexity theory suggest that this kind of hierarchy is ubiquitous in nature. Moreover the grasp of this kind of hierarchy is arguably essential to the perception of things as enduring entities. When we visually perceive an object, especially in the manner of detailed appreciation, what we take in is not a static conglomeration of its many features in various conditions; rather what we perceive is a distinctive pattern of continuity-in-change as these features are transformed through rotations, changes of light, and so forth (cf. Gibson 1973). This dynamic continuity is the object encountered as a distinct value. Is it such a stretch, then, to say that the object's very existence, which is also a kind of continuity through change, is itself an achievement of value?

Of course the kinds of values that are important for an organism to engage are dependent on the interests and purposes of the organism. But given these interests, some structures of the world will serve engagement better than others—especially engagement that is marked by skill.

It is for these reasons that I am pushing for a literal reading of Edward Reed's claim that life is animated by 'the effort after meaning and value' (1996, p. 96): the growth of every living thing is marked by the effort to engage as richly, widely, and deeply as possible with values that matter to its particular way of life. The *purposeful* effort after value may be restricted to a few especially complex creatures, and the enjoyment of value-rich experience for its own sake may be quite rare. But engagement with value is a constant selection pressure for all living things; it may even be the essence of life itself. Life, after all, is not just dependent on a process of value discovery and creation; rather it *is* this process, which can be detected in the most primitive as well as the most sophisticated endeavours. There is a deep connection, in other words, between a bacterium's blind striving for food and Picasso's experimentations with the famous *Las Meninas* of Velázquez.

HUNTER-GATHERER CULTURES AS TECHNOLOGIES OF ENGAGEMENT

At the outset I proposed that religion's origins lie in the various forms of value-rich engagement that constituted distinct ways of life within various ancestral environments. This is not to say that all religious phenomena can be reduced to such forms. As religion gradually distinguished itself it seems also to have disengaged from many basic human pursuits, as well as the wider environmental context of human life. However, insofar as value-rich engagement can occur in any environment, including those dominated by society, the story of religion's development could be told as the shifting of the main locus of engagement rather than disengagement *per se*.

Alas no attempt can be made to tell this story here. I have space (and resources) only to discuss how the turn to value changes our view of one of its earliest and most important chapters. In particular, I wish to show that the tendency to view the cultures of hunter-gatherer peoples in modern religious terms, as systems of belief that give meaning and value to life, entirely misses the more basic importance of value for engagement.

To begin, note that the causal conditions of engagement are a two-way street: just as rich experience depends on a correspondingly rich environment, skill mastery of complex matters requires the development of more refined habits of engagement. Accordingly, for all ways of human life, we should expect complex and vital matters to give rise to technologies of the imagination, that is, to cultural forms that, by specially enhancing the value structure of experience, facilitate deep engagement with relevant values of the environment. In the case of the ways of life dominated by hunting, the critical importance of sensitive and skilful engagement with various animal species has typically given rise to the cultural techniques of engagement known as animism. Here we focus on perhaps its most conspicuous feature, its view of animals as persons.

In his fascinating collection of essays on culture and nature (2000), the anthropologist Tim Ingold mounts a powerful argument for treating animism as a form of skilful engagement. I have just argued that skilful engagement is necessarily also value-rich experience. Accordingly, animistic cultures should evidence both skilful engagement and value-rich experience as two sides of the same coin. And in fact something like value enrichment seems already to be present within Ingold's account. To view animals as persons is not just a way of giving them special value; it is a way of engaging them as complex values in their own right.

While Ingold does not explicitly discuss engagement in terms of value, his readings of animism reveal several telling signs of value-rich experience. For instance, one of the themes of his analysis is his emphasis on perceptual skill rather than propositional knowledge: forms of animistic engagement are traditions of 'enskilment' that, through stories and songs, orient and shape perception; their function is to provide the requisite sensitivity for entering into intricately responsive relationships with animals. Picturing skilful engagement not as a body of propositional knowledge but an acquired ability to 'enter into' relationship signals the importance of depth, a key feature of value-rich experience.

The pronounced value structure of animistic engagement is also indicated by its use of what Ingold calls 'clues': images that 'condense disparate strands of experience into a unifying orientation which, in turn, opens up the world to perception of greater depth and clarity' (2000, p. 22). The various personalities of animals expressed in song, story, and ritual are presumably used as clues in this way. Importantly, such clues do not encode information. Rather, they train attention to appreciate animals as distinctive kinds of value structure: singular perspectives that constitute distinct ways of life in close relationship with the hunter's way of life (2000, p. 51). Personhood, in this context, is not a set of anthropomorphic traits ascribed to animals, but a way of engaging the basic value structure of all living things. The knowledge cultivated by animistic culture is thus 'personal' in the sense that it consists of 'an intimate sensitivity to other ways of being, to the particular movements, habits, and temperaments that reveal each for what it is' (2000, p. 98).

It may be helpful to note the resonance of this sense of 'personal'—a way of entering into relationship rather than the simple ascription of anthropomorphic traits—with Martin Buber's famous distinction between relation of 'I-it' and 'I-You'. Near the beginning of *I and Thou*, Buber describes a 'personal' encounter with a tree: '[I]t can also happen...that as I contemplate the tree I am drawn into a relation, and the tree ceases to be an It...The tree is no impression, no play of my imagination, no aspect of a mood; it confronts me bodily and has to deal with me as I must deal with it—only differently' (1970, p. 58). However, it should also be noted that this value-sensitivity, while conducive to conservationism, is different from the sentimental regard for animals common to modern conservationists. In fact, it invests the hunter-animal relationship with tension, moral ambiguity, and even, at times, existential uncertainty (Ingold 2000, p. 123).

Ingold also affirms the intimate link between value and engagement by denying the separation of aesthetic enjoyment and imaginative play from mundane, everyday subsistence activities. He insists that the widespread production of animal depictions in animistic cultures is not an artistic or religious pursuit separate from hunting (2000, p. 130). A hunter's skill at carving animal figures, for instance, demonstrates how skill mastery and appreciation go hand in hand: hunters are naturally drawn to explore the forms and qualities of the animals they hunt, as they have become connoisseurs in an extraordinarily full sense that encompasses and combines multiple forms of aesthetic delight and skilful interaction. In this context, learning how to track an animal and learning how to evoke its dynamic character in miniature ivory carvings are seamless aspects of intimate involvement with that species.

CONCLUSIONS FOR THE HISTORICAL DEVELOPMENT OF RELIGION

The tandem of value-rich experience and skilful engagement in hunter-gatherer cultures is striking but not at all unique: countless human activities throughout history have achieved the same combination. What is unusual, perhaps, is the comprehensiveness and integrity of such cultures as a way of life. Ingold protests against the tendency of anthropologists to separate these cultures into 'practical-technical interaction with environmental resources in the context of subsistence activities' on the one hand, and 'mytho-religious or cosmological construction of the environment in the context of ritual and ceremony', on the other (2000, p. 56). If he is right, then perhaps religion as a distinct sphere of experience emerged only after the stratification of society and the division of labour enabled the separation of some kinds of value-rich experience from the means of subsistence.

How, then, does the turn to value alter our view of religion's subsequent emergence and development? My sense is that this is an empirical question that can only be answered on a case-by-case basis. The 'effort after value', the basic impulse that drives all human biocultural evolution, is too widespread, too variable, and too fissiparous to support a grand narrative of religious history like those that can be offered from, say, a sociological perspective (e.g. Bellah 2011). In fact the view of religion as a form of value-rich engagement depends on other theoretical approaches in order to distinguish religion from other forms. As an explanatory strategy, then, it cannot stand alone—which is not to say that it cannot make distinctive contributions. In closing, I offer a few sketches of what these contributions might be.

First, the turn to value calls attention to religion in its many popular forms as practised for the sake of direct enjoyment. Regardless of its importance for engagement, value-rich experience can also be its own reward, and there is a lot to be said for religion simply as a way of enriching everyday life. Indeed, if we set aside our modern obsession with belief, it is immediately apparent that the most conspicuous features of popular religious life throughout history are its celebratory aspects: its way of marking major transitions, seasonal changes, and other patterns of life with feasting and festivities, pageantry, dance, music, and so forth. As John Dewey once pointed out, despite our preoccupation with these direct enjoyments, they seldom attract the kind of serious attention given to intellectual and moral endeavours (Dewey 1958, p. 78). Perhaps understanding the importance of value for skilful engagement can help us to appreciate better the pursuit of value for its own sake. At the very least, we need not assume that value-rich experience is simply a way of sugarcoating activities undertaken for utilitarian purposes.

On the other hand, life constantly presents us with pressing demands for skilful engagement, and in such circumstances most forms of value-rich experience will eventually, in one way or another, be pressed into service as forms of engagement. But of course the pursuit of value for its own sake *is* a form of engagement, so what we are really talking about is the relative strength of multiple competing demands for engagement. Insofar as religious forms of engagement are distinguished from others by their comprehensive scope, religion has diversified and evolved through exposure to a very wide and constantly shifting array of demands, one of which is the experience of value itself. Thus adding value to the evolutionary mix provides an element of creative spontaneity that is characteristic of religious innovation and yet missing in so many accounts of religion's development. Value can be a powerful facilitator of rapid cultural evolution.

As an illustration, consider the following scenario. After a distinct form of value-rich engagement emerges in response to demands of life within a particular sociopolitical context, it can sustain itself or even evolve under the power of value-rich experience enjoyed for its own sake, allowing it to detach from the original, wider context in which it arose. All experience is necessarily dependent on *some* context of engagement, but the complex, nested character of life within human society makes these contexts rather slippery. The lack of strict formal correspondence between the values of experience and the environment also facilitates slippage, so that the development of value-rich experience in one context of engagement can easily be 'exapted' by another (Gould and Vrba 1982).

I suggest that something like this process of 'exaptation' lies behind the relatively sudden blossoming of religious thought in ancient China around the time of Confucius. Elaborate forms of ritual and music—paradigmatic technologies of the imagination—had arisen within the sociopolitical context of aristocratic life during the Zhou Dynasty. As the original sociopolitical context under which these forms arose began to crumble, small groups of petty aristocrats—Confucius among them—became passionately invested in their perpetuation, motivated in part, it seems, by the pursuit of value-rich experience for its own sake (Eno 1990). This early Confucian wellspring of value-rich experience was then 'exapted' by subsequent generations of thinkers struggling to come to terms with the social and political turmoil of the Warring States period, thus giving rise to a diverse family of religious philosophies whose massive influence is still felt today.

Last but not least, I would be sorely remiss if I did not mention Robert Neville's theory of what distinguishes religion as a form of value-rich engagement: namely, engagement with ultimate matters, a la Paul Tillich. For a full account and defence of this view, readers are encouraged to turn to Neville's own works (1996; 2006). The only point I wish to make here is that to consider engagement with ultimacy as an evolutionary possibility, we must first explain the development of the kind of comprehensive but intense value-rich experience that could pick up on such profound and subtle kinds of value. I hope to have indicated the first steps in that direction here.

REFERENCES

Atran, S. 2006. The scientific landscape of religion: Evolution, culture, and cognition. In *The Oxford Handbook of Religion and Science*. Eds. P. Clayton and Z. Simpson. New York: Oxford University Press, 407–29.

Barrett, N. F. 2009. The perspectivity of feeling: Process panpsychism and the explanatory gap. *Process Studies 38*(2): 189–206.

——. 2010. Toward an alternative evolutionary theory of religion. *Journal for the American Academy of Religion 78*(3): 583–621.

Bellah, R. N. 2011. *Religion in Human Evolution: From the Paleolithic to the Axial Age*. Cambridge, MA: Harvard University Press.

Buber, M. 1970. *I and Thou*. (W. Kaufman, Trans.). New York: Touchstone.

Chemero, A. 2009. *Radical Embodied Cognitive Science*. Cambridge, MA: MIT Press.

Costall, A. 1999. An iconoclast's triptych: Edward Reed's ecological psychology. *Theory & Psychology 9*(3): 411–16.

Csikszentmihalyi, M. 1990. *Flow: The Psychology of Optimal Experience*. New York: Harper and Row.

Dewey, J. 1934. *Art as Experience*. New York: Perigee Books.

——. 1958. *Experience and Nature*. New York: Dover Publications.

Edelman, G. M. and G. Tononi. 2000. *A Universe of Consciousness: How Matter Becomes Imagination*. New York: Basic Books.

Eno, R. 1990. *The Confucian Creation of Heaven: Philosophy and the Defense of Ritual Mastery*. Albany, NY: SUNY Press.

Gibson, J. J. 1973. On the concept of 'formless invariants' in visual perception. *Leonardo* 6: 43–5.

——. 1986. *The Ecological Approach to Visual Perception*. Hillsdale, NJ: Lawrence Erlbaum Associates.

Gould, S. J. and E. S. Vrba. 1982. Exaptation—a missing term in the science of form. *Paleobiology* 8(1): 4–15.

Ingold, T. 2000. *Perception of the Environment: Essays on Livelihood, Dwelling, and Skill*. New York: Routledge.

Kelso, J. A. S. 1995. *Dynamic Patterns: The Self-Organization of Brain and Behavior*. Cambridge, MA: MIT Press.

Lee, D. N. and P. E. Reddish. 1981. Plummeting gannets: A paradigm of ecological optics. *Nature* 293: 293–4.

Neville, R. C. 1981. *Reconstruction of Thinking*. Albany, NY: SUNY Press.

——. 1989. *Recovery of the Measure*. Albany, NY: SUNY Press.

——. 1995. *Normative Cultures*. Albany, NY: SUNY Press.

——. 1996. *The Truth of Broken Symbols*. Albany, NY: SUNY Press.

——. 2006. *On the Scope and Truth of Theology: Theology as Symbolic Engagement*. New York: T&T Clark.

Reed, E. S. 1996. *Encountering the World: Toward an Ecological Psychology*. New York: Oxford University Press.

Tremlin, T. 2006. *Minds and Gods: The Cognitive Foundations of Religion*. New York: Oxford University Press.

Turner, F. 1992. *Beauty: The Value of Values*. Charlottesville, VA: University of Virginia Press.

Whitehead, A. N. 1967. *Adventures of Ideas*. New York: Free Press.

7

Religion and the Emergence of Differentiated Cognition

FRASER WATTS

Of all the questions relevant to evolutionary theories of religion, two in particular have received most attention. One concerns whether or not religion is adaptive and, if not, how it developed, given the importance of natural selection in evolution. That is an interesting question, and one addressed by Purzycki et al. in this volume, but I will not have anything to say about it here. The other question concerns what evolutionary developments made religion possible. That is the question I will focus on in this chapter. There must have been developments in cognitive capacities (or in the structural relations between those capacities) that facilitated the emergence of religion. The question is what, exactly, those developments were, and how they gave rise to religion.

There is potentially a very broad family of cognitive approaches to psychology generally, and to religion specifically. The cognitive approach in psychology has emerged as a rigorous but versatile mode of theorizing. It is worth emphasizing that cognitive theories of religion do not necessarily emphasize rational, conscious, or intellectual aspects of religion. Cognitive psychology operates with a much broader concept of cognition, including aspects of cognitive processing that are not open to consciousness.

I support the attempt to develop a specifically cognitive theory of religion, but I want to suggest that the particular cognitive approach to the evolution of religion known as the 'cognitive science of religion' (CSR) is in some ways narrower than it needs to be, and that it is worth considering a broader range of cognitive approaches to the evolution of religion. CSR has been set out in slightly different forms, albeit with common features, by people such as Pascal Boyer (1994, 2001), Scott Atran (2002), Justin Barrett (2004), and Todd Tremlin (2006), and admirably represented in this volume by Ilkka Pyysiäinen. I will argue that these are not the only possible cognitive theories of the evolution of religion, and that there is a strong case for considering a broader range of theories.

I will be arguing in this chapter, first, for the potential importance for religion of evolutionary developments in the central cognitive architecture of humans, and suggesting that evolutionary CSR should take these into account, alongside the roles of the specific modules and devices upon which CSR likes to focus. The closest among current CSR theories to the one I will set out here is that of Steven Mithen in *The Prehistory of the Mind* (1996). However, whereas he focuses on the role that the integration of specific modules plays in the evolution of religion, I suggest that consideration should also be given to the role played by increasing cognitive differentiation. Second, I will suggest that there is no good reason to assume that emerging humanity made the kind of intuitive distinction between agents and non-agents that evolutionary CSR often assumes. Both issues have potentially important implications for the cognitive processes involved in the evolution of religion. In relation to both issues, I will argue that evolutionary CSR is unnecessarily committed to speculative assumptions that constrain its theorizing.

THE 'NEW NATURALISM' OF THE COGNITIVE SCIENCE OF RELIGION

Among critical discussions of the CSR project, I would pay particular tribute to the insightful comments made by Barbara Herrnstein Smith (2009) in her *Natural Reflections*. She makes a clear distinction between the general project of offering a naturalistic account of religion, a project that can be traced through Hume, Nietzsche, Edward Taylor, James G. Frazer, Max Weber, Emile Durkheim, and others, and the specific way in which that naturalistic project has been developed by what Smith calls the 'new naturalists'. As she says, 'Clearly one may be sympathetic to the longstanding project of producing naturalistic accounts of religion but sceptical about various ways—past and current—of pursuing that project' (p. 48).

Like Smith, I do not reject the naturalistic approach to religion of people such as Boyer, in the way that, for example, Paul J. Griffiths (2002) does. Griffiths' criticism is based on a rejection of the attempt to take religion out of the context of theology. My position is that naturalistic and theological approaches are not incompatible, and that there is value in looking at religion from both theological and naturalistic perspectives. Even though they make different assumptions there can be a fruitful dialogue between them. I don't accept that the naturalistic approach to religion is the only valid approach, though I do regard it as a valid and worthwhile project.

With Smith, I accept that the cognitive science approach to religion is a particularly fruitful and timely sort of naturalistic approach. The cognitive paradigm in psychology has been enormously fruitful in psychology so far, and has led to significant advances in the understanding of many human phenomena over the last 40 years. It is therefore highly likely that a cognitive approach to understanding religion will also be fruitful. Also, like Smith, I am concerned about the particular assumptions about cognition made in current cognitive theories of religion. I agree with her that the new naturalism has 'promise for affording better understandings of important features of human behavior and culture' (2009, p. 32). However, like her, I am concerned by the 'intellectual confinements represented by some of the project's characteristic methodological, conceptual, and also, in effect, ideological commitments'. Her word 'confinements' seems very apt, as the project is constrained by what seems a rather arbitrary neglect of alternative approaches to the cognitive origins of religion.

In Smith's opinion, the CSR project attempts to understand how religion arose from the operation of what she calls 'the automatic operation of a number of specialized, innate, and universal mental mechanisms' (2009, p. 35). However, the assumption that cognition operates exclusively through such mechanisms is not adequately defended; it is not the only possible kind of cognitive theory. As Smith also says, 'neither the computational-modular model of mind nor the idea of innate automatically triggered mental mechanisms is a foregone conclusion of cognitive science or any other science' (p. 36). As she also points out, Boyer's 'identification of scientific explanation with descriptions of underlying causal mechanisms is questionable' (p. 37).

Smith claims Boyer's position is that, 'we should understand religious ideas . . . and related practices . . . not as more or less functional (or dysfunctional) human responses to recurrent human conditions and experiences but, rather, as effects of the automatic operation of a number of specific, highly specialized, innate and universal mental mechanisms' (2009, p. 35). She is surely right in protesting that that is not the only approach that can properly be called scientific. Various other approaches to the explanation of religion are available. Among those that Boyer rejects, according to Smith, are intellectualist, emotive, irrationalist, and socio-functionalist approaches. It is the rejection of all alternative naturalistic approaches to the explanation of religion that gives CSR its 'confinements', a rejection that often seems to be taken as self-evidently right, rather than being adequately defended.

EVOLUTIONARY DEVELOPMENTS IN THE STRUCTURE OF CENTRAL COGNITION

In this section, the core of this chapter, I mean to explore the possibility that *homo sapiens* developed a new and distinctive central cognitive architecture, and that this played an important role in the cultural explosion generally, and in the evolutionary origin of religion in particular. The development of specific cognitive modules or devices may also have played a role in the evolutionary origin of religion, but there is no reason why it should bear the whole burden of explanation. CSR has in fact shown a strong preference for explanations in terms of specific modules and devices, but there is no reason why explanations in terms of the structure of central processing should be regarded as incompatible with a CSR approach to the origin of religion. Indeed, I suggest that CSR has unnecessarily neglected the role that developments in the structure of central processing might have played.

Modularity of Mind and the Origins of Religion

I will begin with a critical examination of assumptions about modularity in evolutionary psychology and evolutionary CSR. Evolutionary psychology has tended to make strong assumptions about the modularity of mind. Although not all CSR theorists hold an equally strong view of modularity, assumptions about modularity seem to have played a significant role in shaping CSR proposals about the evolution of religion. My immediate focus, then, is on the question of how far the cognitive architecture is organized into specific modules or subsystems and, if so, how these are involved in the evolution of religion.

I want to suggest that current cognitive theories of the evolution of religion often make stronger assumptions about the modularity of mind than are justifiable on the basis of evidence. This constrains the kinds of theories of the evolution of religion that can be offered. Indeed, these assumptions almost dictate what kind of theory can be advanced. The present strong view of mental modularity can be traced to Jerry Fodor's influential book, *The Modularity of Mind* (1983). Its application to evolution was spearheaded by John Tooby and Leda Cosmides (1992). A very strong view of modularity would assume that cognition is composed entirely of specific modules. A more common, somewhat weaker, position would allow that some cognition may be non-modular but, nevertheless, acknowledge that there are good reasons for evolutionary psychology favouring explanations in terms of specific modules. I suggest that there is nothing to prevent CSR from taking the view that some cognition is not organized in terms of modules, and that developments in central, non-modular aspects of cognition may have been important in the evolution of religion.

There is a certain lack of agreement about exactly what a module is, but a workable definition is that it is 'a functionally distinct processing system of the mind, whose operations are at least partly independent of those of others, and whose existence and properties are at least partly dissociable from the others' (Carruthers 2006, p. xii). Much evolutionary psychology has become identified with a strong position on modularity, often known as 'massive modularity'. However, this should not be an article of faith. As Dunbar and Barrett (2007) aptly comment, 'an evolutionary perspective does not of itself commit us to the claim that the mind is entirely organized on modular principles' (p. 5). How modular the mind is should be seen as an empirical question, not a matter of dogma, albeit an empirical question that is not easily answered.

Like almost everyone in contemporary cognitive science, I assume that there are indeed specific cognitive capacities, which may be modularized. As is often said, the mind is somewhat like a Swiss army knife. For example, there may be a 'theory of mind' module, as suggested by Simon Baron-Cohen (1995) and others (e.g. Barrett 2004); a module that specifically allows the prediction of behaviour of other agents. However, beyond the almost universal view that there are at least some specific modules, there is considerable diversity in assumptions.

The evidence is strongest for domain-specific input capacities involved, for example in vision or hearing (see Carruthers 2006), though some assume that there are specific central modules as well. However, the key distinction is probably between those who assume that the mind is composed of nothing but specific modules, and those who are prepared to allow a role for more fluid general-purpose systems in addition to specific modules. I have no problem with the assumption that there are a number of specific modules, and no problem with the assumption that these might include central as well as input/output modules. However, there seems to be neither empirical evidence nor convincing arguments to justify the assumption that the mind is composed of nothing but specific modules.

There is a danger of sliding, without serious argument, from the assumption that there are many specific modules, to the assumption that the mind is composed of nothing but specific modules. That introduces a kind of reductionism—all mental processes are reduced to the functioning of modules. There seems to be no good reason for assuming that central capacities are organized exclusively in a domain-specific, modular way. On the contrary, it is consistent with the available evidence to assume that there is some general, fluid central cognitive capacity.

There are actually quite good reasons for being sceptical of the massive modularity thesis, as David Buller (2005) has argued. One can accept that emerging humanity faced a variety of specific challenges, but it doesn't follow that only specific abilities would be adequate to meet them, or that general problem-solving capacities couldn't have coped. The evolution of specific abilities would have depended both on an organismic structure that enabled specific abilities to emerge, and on an environment that was sufficiently stable for those new abilities to prove their adaptive value. However, it is not clear that the new challenges that faced emerging humanity were actually very stable. As Buller points out, such claims are more a matter of assertion than argument, and don't stand up well to critical scrutiny.

Where you stand on modularity makes a crucial difference to how you can frame cognitive theories of the evolution of religion. It may also be possible to explain the evolutionary origin of religion, and other aspects of the cultural explosion, in terms of the development of central processing capacities. One possible reason why CSR has neglected that possibility is its strong emphasis on modularity of mind (even if the possibility of central processing is not denied). Even if it is allowed that there might be some general, central processing capacity, that does not take the form of specific modules, CSR seems to operate with a scientific paradigm that regards explanations in terms of modules as preferable to explanations in terms of central capacities. That, in turn, may reflect the fact that many of the early CSR theorists were cognitive scientists.

One possible view is that religion is its own special cognitive domain, and that religion arose from the development of a module that was specific for religion. However, the consensus is against the view that there is a module for religion, and I agree with it. For one thing, it is difficult to render this idea plausible in evolutionary terms. It is hard to come up with a convincing account of why a specific religion module would have evolved. The neuropsychological evidence is also increasingly suggesting that neural pathways involved in religion are involved in other things too (Kapogiannis et al. 2009).

Rather than there being a dedicated area of the brain that subserves religion, it seems that religious practices, beliefs, and experiences are so diverse that there is hardly any area of the brain that is not potentially involved in religion in one way or another. Religion seems to be an emergent capacity of the whole brain, rather than having its own dedicated domain-specific capacities, as most of the empirical evidence garnered by CSR's supporters has demonstrated. We have perhaps been over-impressed by the example of language as a specific set of capacities, with its own brain module. It may well be the exception rather than the rule. It seems extremely unlikely that there is a specific module for religion in the same way that there is a specific module for language, or for theory of mind.

If you have ruled out a specific module for religion, and if you have also adopted a strong modularity position that is reluctant to explain the development of new capacities other than in terms of modules, you are left with limited options for explaining how a capacity for religion arose. However, if you allow the possibility that there has always been general-purpose cognition, with central systems handling data from a variety of domains, the possibility opens up that the cognitive capacities of emerging humanity operated rather generally. There may have been developments in these general cognitive capacities that made religion possible.

Differentiation of Central Cognition

I want now to set out a cognitive theory of the evolution of religion, based on different assumptions to the standard CSR model, though not incompatible with it. There is increasing agreement that humans have two cognitive systems, one fast and automatic, the other slow and conscious (e.g. Evans and Frankish 2009; Evans 2010). That is actually quite widely accepted, in some form or other, by CSR theorists. Barrett (2004), for example, makes a distinction between reflective and non-reflective beliefs, the latter being the unconscious intuitive products of a specialized cognitive 'tool-kit'. However, I suggest that the significance for the evolutionary origin of religion of humans having two cognitive subsystems has not been fully recognized by CSR.

On the view that I shall now set out, it is a distinguishing feature of humans that they have two different modes of central cognition, and that this is what made the 'cultural explosion', and ultimately the emergence of religion, possible. In this section of the chapter I will show how an account of the evolution of the distinctive two-system architecture of humans can provide an explanation of the evolution of religion, and other cultural features of life in the Upper Palaeolithic. I will couch this in terms of Philip Barnard's theory of 'interacting cognitive subsystems', or ICS (e.g. Teasdale and Barnard 1993).

As I have argued elsewhere (Watts 2013), there are good reasons, apart from considerations of evolution, for framing cognitive theories of religion in terms of a two-level cognitive architecture. This doesn't necessarily have to be couched in terms of ICS; other two-level or two-system models of the cognitive architecture are available. For example, Hall and Porter (2004) have suggested an approach to religion in terms of Bucci's distinction between symbolic and sub-symbolic cognition (Bucci 1997), and Hill and Hood (1999) have suggested an approach to religion in terms of Epstein's (1994) distinction between an analytical–rational system and an intuitive–experiential system.

Though ICS is not the only two-system approach to religion, it has clear advantages. It is more rigorous, detailed, and explicit in its assumptions than the other two-system models that have been applied to religion; indeed it is probably more rigorous than any other comprehensive model of cognitive architecture. It has also given rise to a very broad range of psychological applications, including psycholinguistics, human-computer interaction, depression, and evolutionary psychology, though it is beyond the scope of this chapter to set out the empirical considerations from those various domains that have led to ICS taking the form it does. ICS has already been applied to religion generally by Clarke (2008), Watts (2002a; 2002b; 2013), Wynn (2005), to mindfulness by Teasdale, (Teasdale et al. 1995; Teasdale and Chaskalson 2011), and to mysticism by Geels (2006).

Barnard (2012) makes a clear and helpful distinction between three psychological levels. Religion is largely manifest at the level of *behaviour.* ICS is concerned with the structure and functioning of the *mental architecture* that underpins that. Mental architecture needs to be distinguished again from the *neural structures* on which it operates. Barnard is a strong advocate of the importance of keeping these three levels distinct, and of the value of mapping behaviour on to an explicit model of the mental architecture before looking at the neural basis. His theory focuses very clearly on the cognitive architecture rather than either of the other levels.

ICS postulates a nine-system mental architecture in humans, but also traces how that was developed through a series of steps in mammalian evolution (see Barnard et al. 2007; Barnard 2010). Barnard suggests that the basic mammalian cognitive architecture had four subsystems, which gradually developed into the nine-subsystem architecture of humans, through a process by which subsystems divided, analogous to cell division in biology. It is proposed that the original four-subsystem architecture consisted of a multi-modal central processor, and three sensory subsystems, concerned with visual, acoustic, and body-state information. The next two subsystems to be added arose from the need for increasingly skilful interaction with the environment, one supporting increased motor dexterity (i.e. 'skeletal effectors'), and the other a more general spatial-praxic system. That gave a mental architecture of six subsystems. The next step supported the increasing development of verbal skills, and involved both a morphonological and an articulatory subsystem. That resulted in an architecture of eight subsystems.

It is the final development of a ninth subsystem which is crucial for religion, and for the 'cultural explosion' in general. This involves the development of a new central subsystem, separate from the previous multi-model subsystem, that for the first time makes specialist provision for propositional semantics. Barnard calls this the 'propositional' subsystem. This new system could not only grasp and encode meanings; for the first time it could propositionalize them. It could also propostionalize emotions, without experiencing them. The propositional subsystem would, for example, be able to propositionalize the distinction between animate and inanimate worlds. It is unnecessary, from an ICS point of view, to postulate separate modules for that. The propositional subsystem is unique in that it interacts exclusively with other central subsystems, not with peripheral ones; that gives it a special role in co-ordinating central information. It is also important that the propositional subsystem has a code that is more closely related to language and can feed into articulations of verbal meanings, something that is distinctive to humans.

The development of this new 'propositional' subsystem, would have had implications in turn for the remainder of what was previously a multi-modal central processor. Barnard calls the other central subsystem the 'implicational' subsystem, a system that is able not only to form meanings arising from interaction with the outside world but also, now, to form schemas concerned with self, others, and the world. There is also a central loop allowing for exchange of information between these two central systems, an exchange that can go on in parallel with engagement with the world through more peripheral subsystems. This creates possibilities of innovation that far exceed what was possible for the previous eight-system architecture that had only one multi-modal, central processor.

The implicational subsystem is schematic, and tends to operate with rather general schemas that have a very broad range of applications. To take a religious example, the 'death to resurrection' schema in Christianity has a very broad range of instantiations, historical, liturgical, moral, and experiential. ICS assumes that the implicational system uses a highly abstract form of encoding, which crosses domains, abstracting the same schemas in a variety of contexts. It is an important strand of ICS theory that cross-domain cognition is characteristic of the implicational system. There is no need to invoke domain-violation to explain cross-domain cognition; it is a characteristic property of implicational cognition. Once the evolution of cognition has given rise to such a subsystem, it is likely that the kind of cross-domain thinking that is characteristic of religion will arise.

It is another important feature of the implicational system that it has no direct route to articulation. Articulation has to be mediated through the propositional system; it cannot be direct. This indicates another sense in which much religious cognition is characteristically implicational. There has been much comment, by William James and others, about the 'ineffability' of mystical experience. It is paradoxical that many mystics comment on the inexpressibility of their experiences, even though they write at length about them. There are various ways of explicating ineffability, but one is in terms of a cognitive architecture, such as that of ICS, which assumes that religious experiences are characteristically encoded in the implicational system, which has no direct route to articulation. ICS would assume that mystics could translate their experiences into the different code of the propositional system, and could then write and talk about them; they are clearly able to do so. However, they also clearly have a sense of loss in doing that. The experiences they are articulating are no longer encoded in their original form.

A third significant feature of the implicational system is that it has input from body states, and has an affective tone. There is, of course, a connection between the two. A distinction has often been made between affective and non-affective cognition. Some time ago, Gordon Bower (1981) distinguished between 'hot' and 'cold' cognition. Implicational cognition is affective. It moves people; it gives rise to what John Wesley memorably called a 'warmed' heart. This is yet another sense in which the assumptions of ICS about the cognitive architecture fit religion very neatly.

The evolutionary division, assumed by ICS, of the central engine of cognition into two separate subsystems—implicational and propositional—would have facilitated the development of religious cognition in several ways. For one thing, it would have freed up the implicational system to be more purely intuitive and schematic than before. That would have facilitated religious cognition for the various reasons given above. Equally, having a propositional system capable of more linguistic thinking would have made it possible to formulate religious assumptions and belief systems in a more explicit way than ever before. After all, having two such systems makes it possible for there to be an interplay between them.

In religion, there is a constant interplay between the more experiential/intuitive aspects of religious cognition and the more explicit and belief-focused aspects. Neither would work well by itself. It is the interaction between the two that enables religion to operate in the way it does. The interplay is well illustrated in the application of ICS to depression (see Teasdale and Barnard 1993). Depressive schemas operate at a non-reflective 'gut' level in the implicational system; that is a significant aspect of what is meant by depressive mood. Those negative schemas give rise, in turn, to potentially negative thoughts that are encoded in the propositional system. Such negative thoughts, in turn, reinforce negative schemas in the implicational system. In religion, there is a similar interplay between intuitive and densely-encoded religious apprehension, and more explicit and potentially articulate religious cognition, in the form of vocal prayer etc. Each gives rise to, and influences the other.

One attraction of this kind of theory is that it provides, at a stroke, an explanation for the various aspects of the cultural explosion that marked the emergence of the human race. A good deal of recent work on the development of religion has focused very specifically on how religious capacities developed, but that leaves us needing a set of similar explanations about how other aspects of the cultural explosion, such as art, developed. It is unparsimonious and unattractive in terms of scientific strategy to advance separate theories for each aspect of the cultural explosion. The kind of central explanation that Philip Barnard is proposing provides a single explanation for all aspects of the cultural explosion that occurred in the Upper Palaeolithic period. In ICS theory, it is the differentiation of central cognition into separate implicational and propositional systems that makes all distinctively human capacities possible. Barnard assumes that it may have been some time before the full potential of this new cognitive architecture was realized. Indeed, there must have been a gradual co-evolution at neural, cognitive, and behavioural levels.

Barnard's theory of cognitive evolution focuses on the cognitive developments that made religion and other aspects of the so-called 'cultural explosion' possible. In other words, it explains how the potential for religion arises. However, the fulfilment of this potential is another matter, and it would not be expected that all the human capacities made possible by the distinctive cognitive architecture of humans became apparent immediately. The capacity for every new capacity made possible by the distinctive cognitive architecture of humans, including religion, would have been realized only slowly. The gradual emergence of religion would have involved processes of cultural evolution, subsequent to the cognitive-neuropsychological evolution that made religion possible. That would involve a gradual process of co-evolution of neural, cognitive, and social structures and functions. The implications of the ICS theory of cognitive evolution for religion, specifically, primarily concern how developments in the cognitive architecture made religion and other cultural developments possible.

Convergent Neuropsychological Proposals

Though Barnard's theory of human evolution and distinctiveness is explicitly a cognitive one, there are also neurological theories of human evolution that point in the same direction. There has been disappointingly little neuropsychological work on human evolution and distinctiveness. For example, there is very little about the brain in the classic book on the origins of modern humans, *The Human Revolution*, edited by Mellers and Stringer (1989), though William Noble and Davidson's (1996) work on the evolution of the human mind pays some attention to the brain. However, there are two recent theories of the neuropsychological evolution of humans that are consistent with the claim that humans are distinctive in having two separate modalities of central cognition.

One is the theory developed by Thomas Wynn and Frederick Coolidge (2012), which focuses on the distinctive oppositional processes of the human brain, one of which is the opposition between the frontal lobes, concerned with planning and higher-order reflective schemas, and the older brain of the temporal and other lobes. Wynn and Coolidge emphasize the way in which humans have augmented the more primitive capacity for unplanned, spontaneously adaptive activity, found in all primates, with the distinctive capacity for planning made possible by the exceptionally well-developed frontal lobes of humans. Amati and Shallice (2007) take a similar view about the distinctive human capacity for the planning and regulation of behaviour. It is perfectly possible that the novel human capacity for planning leads to two modes of cognitive organization of activity, planned and unplanned, which co-exist and operate in tandem.

Even more relevant to the two-system theory of human cognition is lateralization theory, which focuses on another basic oppositional process in the brain—the contrast between left and right sides of the brain. One of the intriguing aspects of lateralization, as McGilchrist (2009) has recently pointed out, is that the two halves of the brain are both more different from each other in humans than in our primate ancestors, and also less interconnected with each other. That allows the two halves of the brain to operate in different ways, and relatively independently. This differentiation of the physical brain into two increasingly distinct hemispheres could have facilitated the differentiation of the cognitive architecture into two separate subsystems. In most people, language is subserved by the left side, whereas the right hemisphere subserves a relatively non-verbal mode of cognition.

Things are, of course, not always so simple, and it should be noted that 'the pattern of asymmetries varies with handedness, gender and age, and with a variety of genetic factors and hormonal influences' (Toga and Thompson 2003). McGilchrist is well aware that in the past there has been a lot of over-simplified thinking about 'left brain' and 'right brain', and he is at pains to avoid the idea that one side of the brain does one set of things, and the other side of the brain does another. He means rather to emphasize that both hemispheres can do almost everything in their own way, but that they have identifiably different ways of doing them. He assembles a formidable body of research data in support of this more subtle position. He also recognizes, following Kinsbourne (1978) and others, that the contrast between left and right brain is only one of three opponent processes in the brain, the others being between cortex and subcortex (up/down), and between the frontal lobes and the more posterior cortex (front/back). Savage (2013) has applied all three opponent processes to religion.

We are still developing a complete and integrated neuropsychology of human evolution and distinctiveness, but the works of Wynn and Coolidge, and of McGilchrist are likely to be important building blocks. That integrated theory, when it emerges, seems likely to shed light on why and how humans have two distinct cognitive subsystems.

Developments in Central Processing Capacities

I will now compare the theory of the evolutionary origin of religion that I have set out above, which draws on Barnard's general theory of cognitive evolution, with the CSR theory that is closest to it, in terms of focusing on the evolutionary development of the cognitive architecture. That is the cognitive theory of the origin of religion proposed by Steven Mithen in *The Prehistory of the Mind* (Mithen 1996). There is a convenient summary of it by Wentzel van Huyssteen (2006) in his *Alone in the World?* In essence, Mithen proposes that religion arises from a fusion of technical, natural history and social intelligences.

Mithen makes good use, especially in his more recent publications on religion (e.g. Mithen 2008, 2009), of Harvey Whitehouse's (2004) distinction between imagistic and doctrinal modes of religiosity. It seems to have been 'imagistic' religion (involving infrequent but emotionally-arousing rituals that lead to the formation of tight-knit social groups) that emerges as part of the cultural explosion in the Upper Palaeolithic period. In contrast, 'doctrinal' religion (involving more frequent but tamer rituals in a larger and more diffuse social group) is almost certainly a later development and probably comes with stable human settlements. Bringing this distinction between different modes of religion into discussion about the evolution of religion makes Mithen's approach more sophisticated about the nature of religion than many others. Much CSR theory operates with 'religion' as a rather undifferentiated category.

Another attractive feature of Mithen's approach is that he takes care over the archaeological evidence. Too much evolutionary CSR theorizing glosses over that, and focuses instead on developmental or anthropological evidence from humanity today. While it is arguable that such evidence is relevant to the evolution of religion, it is dangerous to depend on it entirely. Mithen builds on Renfrew's (1985) very careful discussion of the issues that arise in interpreting archaeological evidence for the evolution of religion.

There are basically three approaches to the explanation of the set of functional and behavioural changes that took place in the so-called cultural explosion, and it will be helpful to locate Mithen's approach within this range. One is to have a separate cognitive theory of each individual feature of the cultural explosion, without reference to any other feature. For example, one can advance quite separate theories of the evolutionary origin of religion and art. However, such an approach is unparsimonious; it is also rather implausible to suggest that each feature of the cultural explosion arose for a distinct reason, unconnected with the others. It is one of the very odd features of the CSR approach to the evolution of religion that it has tended to focus on religion alone, taking religion out of the context of other features of the cultural explosion. That is not the path Mithen takes.

In contrast, Mithen sees the task as one of explaining the cultural explosion as a whole, including religion, rather than of explaining the evolution of religion in isolation. One way of approaching that is to take one of the many changes that arose at the dawn of the Upper Palaeolithic period, and to argue that it is **the** fundamental change from which all others flow. Mithen (1996, p. 174) mentions several such theories. These include the restructuring of social relations, the appearance of economic specialization, and the origin of language. Other such theories could be advanced (Amati and Shallice 2007). Mithen is also unenthusiastic about this kind of search for a single key functional change, from which all others are thought to flow.

Mithen suggests that a better approach to the explanation of the cultural explosion is to focus on the underlying structural changes in the cognitive architecture from which this wide range of new cognitive capacities flowed. This seems to me an attractive approach in terms of general scientific strategy. Up to this point, I am in complete agreement with his approach. However, I suggest that he is unconvincing when it comes to suggesting exactly what the key change in the cognitive architecture is. The theory that I have set out above, in which I have applied to religion Barnard's ICS theory of cognitive evolution, takes a different path.

There are certainly commonalities between Mithen's theory and the theory I have derived from Barnard. Both take as their starting point some relatively undifferentiated processing capacity, and propose a process of differentiation. However, for Mithen, that differentiation involves the development of specific modules (a natural history intelligence, a social intelligence, and a technical intelligence), whereas, on the theory that I have derived from Barnard, the key change is a differentiation of a single central processor into two more specific central subsystems.

Both Mithen and Barnard then envisage increased integration between the components of their more differentiated cognitive architectures. For Mithen, that takes the form of dissolving the boundaries between the specific modules that have been developed, in order to create fluid intelligence. In contrast, for Barnard, the differentiation between two distinct central modules is maintained, though there is increasingly effective cross-talk between them.

Finally, there is a key difference between Mithen and Barnard in where they put the emphasis in explaining the cultural explosion. For Barnard, the crucial step is the differentiation of central processing into two distinct subsystems. In contrast, for Mithen, the crucial step is the dissolution of boundaries between specific modules. The basic contrast is thus between Mithen's increasing cognitive *integration* or Barnard's increasing cognitive *differentiation* (even though both Barnard and Mithen would allow that there is an element of both).

At very least, it does not seem necessary to propose, as Mithen does, that the boundaries between specialized intelligences (e.g. technical, natural history, and social intelligences) were abolished. Rather, it seems at least as likely that specialized intelligences remained separate, but become capable of working together in a more integrated way. There could well have been *both* a tendency for cognitive capacities to become more differentiated and specialized, and *also* a capacity for those specialized intelligences to work together in a more integrated way. The new capacity for religion, art, and other aspects of the cultural explosion may have depended on both evolutionary trends.

Mithen, to his credit, is fully aware of the reasons why one might emphasize that humans came to have a more differentiated, rather than a more integrated, cognitive architecture. However, in the end (for no good reason, as it seems to me) he fails to be persuaded by arguments that he has himself set out for the importance of increased cognitive differentiation, and opts for cognitive integration as the key feature of emerging humanity. Nevertheless, and very commendably, he is open-minded enough to realize that there is a significant body of developmental work that does not fit this approach and, for example, he cites Patricia Greenfield's argument that, prior to modularization, different skills rely on the **same** cognitive capacities.

The structure of Mithen's theory is in some ways reminiscent of Julian Jaynes' (1976) book on the evolution of consciousness, which argued that the new kind of human consciousness associated with religion (especially hearing the voice of God) arose from the breakdown of the 'bicameral' mind, i.e. from closer integration between the hemispheres. Jaynes' argument was that in emerging humanity the two hemispheres of the brain worked fairly independently, and that the crucial development for the origin of religion was the breakdown of this compartmentalization, so that the two hemispheres could work in a more integrated way. Hearing the voice of God was essentially one hemisphere 'hearing' the other. It is a story of the breakdown of divisions between different elements in the cognitive architecture that is in some ways analogous to that of Mithen.

Iain McGilchrist (2009) has recently argued, as part of an extended and persuasive treatment of hemispheric lateralization that Jaynes got things the wrong way round, and that the phenomena that Jaynes attributes to the breakdown of the bicameral mind actually came about because of the development of the 'relative separation of the two chambers' (p. 262). The two hemispheres have more distinct functions in modern humans than in our primate ancestors, and there is less anatomical connection between the hemispheres in humans than any other related species. As far as the evolution of the brain is concerned, the trend is towards greater differentiation, rather than to the breakdown of cognitive divisions. This very significant development in the brains of *homo sapiens* towards greater hemispheric specialization and distinctness could well have been the key change that underpinned the development of language, art, religion, and other aspects of the changes that occurred in the Upper Palaeolithic period.

Of course, it is not being suggested that the two hemispheres operate independently in humans, but it seems that the crucial change may have been their becoming more distinct before integration could be achieved. Mithen puts the emphasis on the breakdown of modular capacities as the crucial change in cognitive capacities that underpinned the Upper Palaeolithic changes. In contrast, the crucial precondition could equally well have been the prior development of a more differentiated central cognitive architecture, which would have been facilitated by greater separation between the hemispheres. That would have led to increasing interchange and cross-talk between these distinct central subsystems, producing the increased cognitive fluidity that is the key feature of Mithen's proposal.

ONTOLOGICAL CATEGORIES: AGENTS AND NON-AGENTS

I will now turn from the differentiation of the cognitive architecture to the differentiation of the perceptual world of emerging humanity. Having raised the possibility, following Barnard, that the central cognitive architecture may have become more differentiated with the emergence of humanity, I will also raise the possibility that the perceptual world of emerging humanity may gradually have become more differentiated. CSR theory often implies that the distinction between agents and non-agents was intuitively obvious for emerging humanity. However, that distinction may actually have been acquired only slowly and gradually. That leads to a somewhat different proposal from those so far advanced in evolutionary CSR about why and how early humanity had a propensity for animistic cognition. If non-agents were often perceived as agents, that could have been because the distinction between agents and non-agents was not yet intuitive, but was still being developed.

Agents and Non-Agents in Cognitive Science of Religion

Many CSR theorists assume that the world divides up into different ontological categories (e.g. Tremlin 2006). This division of the world into ontological categories is somewhat parallel to the division of the mind into modules, and may even be what drives the division of the mind into modules. For each ontological category there is a set of intuitive rules, expectations, and assumptions about how members of that category function.

CSR makes an important distinction between agents and non-agents. From a standard CSR point of view, the puzzle about religious cognition (and indeed about the broader category of animistic thinking), is that it applies to non-agents a way of thinking that is only appropriate to agents. This is the core of Guthrie's (1993) proposal for the origin of animistic thinking, and it has been developed in slightly different, albeit overlapping ways, by different CSR theorists. I will focus on two of the main ways in which it has been developed: 'minimal counterintuitiveness' (Boyer 1994) and 'hyper-sensitive agency detection' (Barrett 2004).

Boyer's proposal is that religious ideas partly fulfil intuitive assumptions, but partly diverge from them. As he says, 'religious representations are particular combinations of mental representation that satisfy two conditions. First, the religious concepts *violate* certain expectations from ontological categories. Second they *preserve* other expectations' (2001, p. 71). Religious representations are counterintuitive because they break intuitive assumptions.

However, he proposes that religious concepts break intuitive assumptions only minimally. 'Minimally counterintuitive' representations are held to be especially memorable.

It is certainly a well-established fact in social cognition that schema-incongruent information (i.e. information that relates to a schema but violates schematic assumptions) is more memorable than either schema-congruent information or a-schematic information (i.e. information that departs from schemas so radically as not to connect with them). So, there is no problem about the assumption that 'minimally counterintuitive' information would be memorable. However, it is not clear that the memorability of 'minimally counterintuitive' information necessarily plays the key role in the evolution of religion that Boyer thinks it does. The key question is whether the ideas of emerging humanity about supernatural agents were actually, for them, 'counterintuitive'. It seems to me that we just don't know enough about the intuitions of emerging humanity to be sure of that; it is too easy to assume that emerging humanity had the same intuitions that we do.

In CSR studies of how counterintuitive information is processed, judgements of what is counterintuitive are often those of contemporary CSR researchers, most of whom are naturalists. My point is that what is counterintuitive for contemporary naturalists may not have been counterintuitive for emerging humanity. The distinction between naturalism and supernaturalism that seems so obvious to us may not have been obvious at all for early *homo sapiens*. The distinction between the ontological categories of agents and non-agents may have been a hard-won cognitive achievement that was reached only gradually. Before that conceptual distinction was achieved, it may not have been counterintuitive to think in animistic or religious ways.

Barrett (2004) and others have proposed a hypersensitive agency detection device (HADD) that detects agency where it should not be detected. Barrett (p. 33) says that agency can be attributed to something ambiguous such as 'a wispy form'. Alternatively, it may mistakenly be imagined that 'known non-agents are exhibiting agency' (p. 33). For example a cloud can be taken as an agent. Cognitive interpretations that belong to one domain can be misapplied in another (i.e. agency cognition can be misapplied to non-agents), producing what is sometimes called 'domain violation'. It is argued that there were survival advantages for emerging humanity, living by hunting and perhaps also being hunted, in detecting agency wherever it might arise, and that it was better to detect it too often than too infrequently. Again, it is not obvious that the categories of agents and non-agents were separate and distinct before the functioning of HADD gave rise to supernatural thinking and domain violation. To put it another way, it is hard to be sure that what Barrett calls 'known non-agents' were actually known by emerging humanity to be such.

Typically ECSR's assumption seems to be that there was a clear, intuitive, distinction between agents and non-agents, but that HADD led to agency thinking being extended to non-agents. It is equally possible that there was originally no clear ontological distinction between agents and non-agents, with the result that many non-agents were originally treated as agents. Gradually, emerging humanity may have begun to apply the intuitive rules of agency detection more narrowly, until they were limited to the identification of what would now be recognized as agents. Of course this is an empirical matter, but one that we just don't have the evolutionary data to settle. My point is that the HADD story is not obviously correct; at very least, it could have been otherwise.

Justin Barrett's excellent programme of research on the cognitive development of religion in children (see, for example, Barrett 2012) actually fits very well with the suggestion that agency detection developed from initially being undifferentiated to becoming more differentiated. His research shows that young children, at about age three, have a single category of agents. They make no distinction between supernatural and human agents. That distinction comes later. They do not start with a concept of human agents and then expand it to include supernatural agents. They start with a single category of agents, assuming that all agents are like gods. Gradually, they learn the distinction between gods and humans, and that humans are more limited than gods. Human development is, of course, not evolution; ontogeny does not necessarily recapitulate phylogeny. However, Barrett's developmental data suggests a revision to evolutionary CSR theory that emphasizes increasing differentiation between categories rather than a loss of differentiation between them.

It may be helpful to link the question of the relationship between agency and non-agency thinking in emerging humanity to a parallel debate about the development of the meanings of words, especially the relationship between words for material objects and for psycho-spiritual experiences. Similar issues arise, and they may come into focus more clearly from exploring the parallel. If you trace back the meanings of words, you find a tendency for the meanings of words to be double-aspect ones, spanning material and experiential domains (Barfield 1928). The Hebrew *ruach*, breath and spirit, is a classic double-aspect term. There is no evidence for the common assumption that words were originally literal in their meanings, and then became metaphorical. In fact it seems that they began by having double-aspect meanings, from which more specific meanings were distilled (see Watts and Williams 1988). It is interesting that the social psychologist, Solomon Asch (1958), found the same double aspect meanings (i.e. the same linkages between meanings in the material and psycho-social domains) in historically independent linguistic traditions. So, for example, you find the same double-aspect linkages for both sweet and bitter in Hebrew, Greek, Chinese, Thai, Burmese, and Hausa, i.e. historically independent languages.

It is often assumed that words must have begun with literal meanings, and in particular that they initially referred to concrete, material entities. It is assumed that they were subsequently used metaphorically, to refer to other aspects of subjective experience. However, there is another equally plausible possibility, that words began by being domain-general, referring to related meanings in different domains, and subsequently became more refined in their meanings, as they came to be used more literally. Just as emerging humanity may have made no clear distinction between agents and non-agents, so there may initially have been no clear distinction between words for material and psycho-spiritual realities. In both cases, I suggest, it is at least possible that the world was perceived in a relatively undifferentiated way by emerging humanity.

Metaphysical Basis and Theological Significance

Finally, I will draw out the metaphysical background and theological implications of this theory more explicitly and, in particular, examine how it relates to theistic or non-theistic assumptions. Which choice one makes in framing scientific theories is often influenced by background metaphysical assumptions, and they are likely to be relevant to whether people prefer the view that there was initially a clear, intuitive distinction between agents and non-agents that was somewhat obscured by HADD, or the view that the intuitive distinction between agents and non-agents emerged only gradually. The latter view, if it turns out to be correct, would be more congenial from a religious point of view, because it avoids the assumption that religion is some kind of cognitive mistake. However, there are motivations on the other side too. Many of those advocating the CSR story, with the notable exception of Justin Barrett (2004), make materialist, naturalist assumptions, from which it is congenial to find support for the assumption that religion is a cognitive mistake.

On the domain violation view, religion seems to be something of a mistake, an inappropriate application of one way of thinking, which is appropriate for a particular domain, to objects and events that properly belong to another. It may be, in some sense 'natural', and indeed memorable, something that may have had advantages for survival, but nevertheless it is a mistake. There is a paradox in seeing religion as, in some sense, natural but also as representing domain violation. In contrast, I suggest that religion might really be natural, not in the sense of being a mistake that it was natural to make, but natural in the sense of being intuitively correct. As I have already pointed out, this possibility fits Barrett's (2012) finding that children are born with an instinctive capacity to think about gods.

Emerging humanity may have seen nature as inhabited by spirit, and seen the heavens as revealing the wrath of God (as in Psalm 18), or declaring his glory (as in Psalm 19), because they intuitively felt it to be so. It may not have been, for them, 'counterintuitive'. Humanity may have increasingly come to be conscious of distinctions between different ontological domains, which would in turn have led to changing intuitions. It could then have become as intuitively obvious that different domains were separate as it had previously been intuitively obvious that they were connected.

It is possible that the intuitive theism of the children that Barrett studied (and the parallel animism and supernaturalism of evolving humanity), might actually be correct. That is a possibility that is obviously more congenial from a religious point of view, just as the standard CSR view that religion is some kind of cognitive mistake is more congenial for those who make naturalistic assumptions. There are admittedly no strong arguments in favour of this possibility, but equally there are no strong arguments for why the naturalistic assumptions of the standard CSR approach should be preferred.

The story of Adam and Eve in Genesis 3 seems to capture, in a mythological way, something important about cognitive evolution. It seems to be about the emergence of conceptual distinctions, explicitly about the distinction between good and evil, and implicitly about the emergence of a conceptual distinction between God and humanity (Watts 2002a, pp. 113–132). At the start of the story, Adam and Eve don't seem aware of such distinctions. However, by the end of the story there is a clearer apprehension of the distinction between God and humanity, which carries both gains and losses. If emerging humanity gradually learned conceptual distinctions between agents and non-agents, and between material and spiritual domains, those distinctions would parallel those learned by Adam and Eve in the mythological story.

It seems to me that the cognitive science of religion has, in effect, been developing a competing myth of a 'fall' from naturalism. There is clearly a possibility of contemporary scientists, who hold naturalistic assumptions, projecting those assumptions back onto emerging humanity. It can seem obvious to them that the world really is purely material, and that emerging humanity must have seen it in that way. Then, it is reasonable to assume that something went wrong and disrupted that correct intuition that the 'natural' world was indeed purely natural. That is the point at which the mistake of extending agency detection to the inanimate world arises. It becomes a kind of alternative 'fall'. However, rather than being a fall from God and paradise, as in Genesis 3, it is a fall from naturalism into religion.

My point, throughout this chapter is that the standard CSR view doesn't have to be correct. There are plausible cognitive alternatives that have not yet been adequately considered. Developments in central cognition might have played the key role in the evolution of religion, rather than the development of specific modules and devices. The relevant changes in central cognition may have been towards a more differentiated cognitive architecture, as Barnard proposes, rather than towards the breakdown of boundaries between separate cognitive capacities as Mithen assumes. Then, just as the cognitive architecture may have become more differentiated, so the perceived world of emerging humanity may gradually have become more differentiated. The tendency of emerging humanity to attribute agency to what we regard as non-agents could simply reflect the fact that the conceptual distinction between the two domains was not yet fully developed. Finally, the possibility cannot even be ruled out that emerging humanity was correct in not embracing the assumptions of contemporary naturalism. The possibility set out here, that the evolution of religion was linked to increasing differentiation in both the central cognitive architecture and the perceived world of emerging humanity, deserves consideration as a variation on the standard CSR story.

REFERENCES

Amati, D. and T. Shallice. 2007. On the emergence of modern humans. *Cognition 103*: 358–85.

Asch, S. E. 1958 The metaphor: A psychological inquiry. In *Person Perception and Interpersonal Behaviour*. Eds. R. Taguiri and L. Petrullo. Stanford: Stanford University Press, 86–94.

Atran, S. 2002. *In Gods We Trust: the Evolutionary Landscape of Religion*. New York: Oxford University Press.

Barfield, O. 1928. *Poetic Diction (1st edition)*. London: Faber.

Barnard, P. J., D. J. Duke, R.W. Byrne, and I. Davidson. 2007. Differentiation in cognitive and emotional meanings: an evolutionary analysis. *Cognition and Emotion 21*(6): 1155–83.

Barnard, P. 2010. Current developments in inferring cognitive capabilities from the archaeological traces left by stone tools: Caught between a rock and a hard inference. In *Stone Tools and the Evolution of Human Cognition*. Eds. A. Nowell and I. Davidson. Boulder, CO: Colorado University Press, 207–26.

——. 2012. What do we mean by the meanings of music? *Empirical Musicology Review 7*(1–2), 69–80.

Baron-Cohen, S. 1995. *Mindblindness: An Essay on Autism and Theory of Mind*. Cambridge, MA: MIT Press.

Barrett, J. L. 2004. *Why Would Anyone Believe in God?* Lanham, MD: AltaMira Press.

——. 2012. *Born Believers: the Science of Children's Religious Belief*. New York: Free Press.

Bower, G. H. 1981. Mood and memory. *The American Psychologist 36*: 129–48.

Boyer, P. 1994. *The Naturalness of Religious Ideas: a Cognitive Theory of Religion*. Berkley and Los Angeles, CA: University of California Press.

——. 2001. *Religion Explained: the Human Instincts that Fashion Gods, Spirits and Ancestors*. London: Vintage Books.

Bucci, W. 1997. *Psychoanalysis and Cognitive Science: a Multiple Code Theory*. New York: Guilford Press.

Buller, D. J. 2005. *Adapting Minds: Evolutionary Psychology and the Persistent Quest for Human Nature*. Cambridge, MA: MIT Press.

Carruthers, P. 2006. *The Architecture of the Mind: Massive Modularity and the Flexibility of Thought*. Oxford: Clarendon Press.

Clarke, I. 2008. *Madness, Mystery and the Survival of God*. Winchester: John Hunt.

Dunbar, R. I. M. and L. Barrett. Eds. 2007. *The Oxford Handbook of Evolutionary Psychology*. Oxford: Oxford University Press.

Epstein, S. 1994. Integration of the cognitive and the psychodynamic unconscious. *American Psychologist 49*: 709–24.

Evans, J. 2010. *Thinking Twice: Two Minds in One Brain*. Oxford: Oxford University Press.

Evans, J. and K. Frankish. Eds. 2009. *In Two Minds: Dual Processes and Beyond*. Oxford: Oxford University Press.

Fodor, J. 1983. *The Modularity of Mind*. Cambridge, MA: MIT Press.

Geels, A. 2006. The Marriage between ego and ID: Cognitive integration and its relation to mystical experience. *Archive for the Psychology of Religion 28*: 219–52.

Griffiths, P. J. 2002. Faith seeking explanation. Review of *Religion Explained*, by Pascal Boyer. *First Things 119*: 53–7.

Guthrie, S. E. 1993. *Faces in the Clouds: A New Theory of Religion*. New York: Oxford University Press.

Hall, T. W and S. Porter. 2004. Referential integration: an emotional information processing perspective on the process of integration. *Journal of Psychology & Theology 32*(3): 167–80.

Hill, P. C. and R. W. Hood.1999. Affect, religion and unconscious process. *Journal of Personality 67*: 1015–46.

Jaynes, J. 1976. *The Origin of Consciousness in the Breakdown of the Bicameral Mind*. Boston, MA: Houghton Mifflin.

Kapogiannis, D., A. K. Barbey, M. Su, G. Zamboni, F. Krueger, and J. Grafman. 2009. Cognitive and neural foundations of religious belief. *Proc. Natl. Acad. Sci. USA 106*(12): 4876–81.

Kinsbourne, M. Ed. 1978. *Asymmetrical Function of the Brain*. Cambridge: Cambridge University Press.

McGilchrist, I. 2009. *The Master and his Emissary: The Divided Brain and the Making of the Western World*. New Haven: Yale University Press.

Mellers, P. A. and C. Stringer. Eds. 1989. *The Human Revolution. Behavioural and Biological Perspectives in the Origins of Modern Humans*. Edinburgh: Edinburgh University Press.

Mithen, S. 1996. *The Prehistory of the Mind*. London: Phoenix.

——. 2008. Is religion inevitable? An archaeologist's view from the past. In *The Edge of Reason? Science and Religion in Modern Society*. Ed. Alex Bentley. London: Continuum International Publishing Group, 82–94.

——. 2009. The prehistory of the religious mind. In *Theology, Evolution and the Mind*. Ed. Neil Spurway. Newcastle upon Tyne: Cambridge Scholars Publishing, 10–30.

Noble, W. and I. Davidson. 1996. *Human Evolution, Language and Mind: a Psychological and Archaeological Inquiry*. Cambridge: Cambridge University Press.

Renfrew, A. C. Ed. 1985. *The Archaeology of Cult, the Sanctuary at Phylakopi*. London: British School at Athens and Thames and Hudson.

Savage, S. 2013. Head and heart in preventing religious radicalisation. In *Head & Heart: Perspectives from Religion & Psychology*. Eds. F. Watts and G. Dumbreck. Philadelphia: Templeton Press, 157–93.

Smith, B. Herrnstein. 2009. *Natural Reflections: Human Cognition at the Nexus of Science and Religion*. New Haven, Conn: Yale University Press.

Teasdale, J. D. and P. J. Barnard. 1993. *Affect, Cognition and Change: Re-Modelling Depressive Thought*. Hove: Lawrence Erlbaum Associates.

Teasdale, J. D. and M. Chaskalson. 2011. How does mindfulness transform suffering? II: The Transformation of Dukkha. *Contemporary Buddhism 12*: 103–24.

Teasdale, J. D., Z. V. Segal, and J. M. G. Williams. 1995. How does cognitive therapy prevent depressive relapse and why should attentional control (mindfulness) training help? *Behaviour Research and Therapy* 33: 25–39.

Toga, A and P. M. Thompson. 2003. Mapping brain asymmetry. *Nature Reviews Neuroscience* 4: 37–48.

Tooby, J. and L. Cosmides. 1992. The psychological foundations of culture. In *The Adapted Mind*. Eds. J. H. Barkow, L. Cosmides and J. Tooby. New York: Oxford University Press, 9–136.

Tremlin, T. 2006. *Minds and Gods: the Cognitive Foundations of Religion*. New York: Oxford University Press.

Van Huysteen and J. Wentzel. 2006. *Alone in the World? Human Uniqueness in Science and Theology*. Grand Rapids, MI: Eerdmans Publishing Company.

Watts, F. N. and M. Williams. 1988. *The Psychology of Religious Knowing*. Cambridge: Cambridge University Press.

Watts, F. N. 2002a. *Theology and Psychology*. Baskingstoke: Ashgate.

——. 2002b. Interacting cognitive subsystems and religious meanings. In *Neurotheology: Brain, Science, Spirituality, Religious Experience*. Ed. R. Joseph. San Jose: California University Press, 183–8.

——. 2013. Dual system theories of religious cognition. In *Head & Heart: Perspectives from Religion & Psychology*. Eds. F. Watts and G. Dumbreck. Philadelphia: Templeton Press, 125–54.

Whitehouse, H. 2004. *Modes of Religiosity: A Cognitive Theory of Religious Transmission*. Walnut Creek, CA: Altamira Press.

Wynn, M. 2005. *Emotional Experience and Religious Understanding*. Cambridge: Cambridge University Press.

Wynn, T. and F. L. Coolidge. 2012. *How to Think Like a Neandertal*. New York: Oxford University Press.

8

From Empathy to Embodied Faith? Interdisciplinary Perspectives on the Evolution of Religion

J. WENTZEL VAN HUYSSTEEN

In some of my earlier work I have argued that a postfoundationalist notion of rationality points beyond the boundaries of our own disciplines, our local communities, groups, or cultures, toward plausible forms of *interdisciplinary dialogue* (cf. van Huyssteen 1999, 2006). Against this background I have argued for distinct and important differences between reasoning strategies used by theologians and scientists. I have also argued, however, that some important shared rational resources may be identified for these very different cognitive domains of our mental lives (2006, pp. 1–44). It is precisely these shared interests and overlapping rational resources that enable interdisciplinary dialogue, and are expressed most clearly by the notion of *transversal rationality*. In the dialogue between theology and other disciplines, transversal reasoning promotes different, non-hierarchical but equally legitimate ways of viewing specific topics, problems, traditions, or disciplines. It creates the kind of space where different voices need not always be in contradiction, or in danger of assimilating one another, but are in fact dynamically interactive with one another. This notion of transversality can enable us to identify possible shared conceptual problems as we negotiate the porous boundaries of our different disciplines.

One such shared cross-disciplinary problem is the concern for 'what makes us human', for human species specificity, and how that may relate to human origins and, possibly, to the evolution of religious awareness. The underlying argument of this chapter is that in the search for the nature of the human self, of personhood, theology and the sciences may find a surprising overlapping, shared research trajectory. However, as we all know today, no one trait or accomplishment should ever be taken as the single defining characteristic of what it means to be human. I will argue in this chapter that it is precisely in the interdisciplinary forum that we may find exciting answers to questions like 'what makes us persons?', and 'what defines the human self?', answers that might ultimately also guide us through the complexity of the current discussion on the evolution of religion and religious behaviour.

Against this background my thesis will be that the question of the evolution of religion and religious behaviour, the central theme for this anthology, can never be disentangled from the evolution of embodied human personhood. This will enable us to evaluate contemporary proposals for aspects of human personhood that were all of great importance to Darwin: the evolution of empathy and attachment, the evolution of morality, and the evolution of the religious disposition, to name just a few. Finally, I will argue that each of these deeply human traits has consistently played a defining role in the evolution of human communication and human interpersonal attachment, and, along with the evolution of complex symbolic behaviour, they combine to give us important insights into the evolution of religion and religious behaviour.

COGNITIVE SCIENCE OF RELIGION AND THE EVOLUTION OF RELIGION

Without any doubt the cognitive science of religion is the dominating voice in current discussions about the origins and evolution of religion. The cognitive science of religion (CSR) is an interdisciplinary research programme that includes the fields of the cognitive sciences (cognitive linguistics, cognitive psychology, cognitive neuroscience, artificial intelligence studies, and cognitive anthropology) and the study of religion. An excellent guide through the vast amount of literature generated by a virtual explosion of literature in this field is given in Aku Visala's *Naturalism, Theism and the Cognitive Study of Religion. Religion Explained?* (2011). Visala argues correctly that central to this research programme over the last twenty years has been the application of theories from both the cognitive sciences and evolutionary epistemology in order to explain the general forms of religious ideas and behaviours (pp. 13–5). Within the larger and loosely integrated fields of cognitive science and evolutionary psychology, there has emerged a relatively tight-knit group of scholars engaged in what now is known as 'cognitive science of religion' (p. 4). In this particular book Visala discusses and critically analyses the work of four prominent members of this group, namely Pascal Boyer, Scott Atran, Justin Barrett, and Dan Sperber.

One of the founding fathers of cognitive science is Noam Chomsky, who claimed that language acquisition and structures of human languages are strongly constrained by innate, psychological capacities of language processing. In this sense, according to Visala, Chomsky was the *first* to formulate the basic idea of 'innateness' or 'nativism' (cf. Visala 2011, p. 11). Here the human mind is not a 'blank slate' which just passively records and memorizes information presented to it. On the contrary, the human mind comes with intuitive, innate biases, schemas, models, and mechanisms that actively shape the acquisition and transmission of information.

The *second* major development in CSR was the emerging consensus from cognitive psychology according to which the innate biases and information processing tendencies of the human mind have different effects in different domains of knowledge. The basic idea of this 'domain-specificity hypothesis' is that in the very early stages of human development we can already see very specialized and context-sensitive cognitive mechanisms at work. (cf. Visala 2011, p. 6).

The *third* essential development contributing to the birth of CSR was the emergence of cognitive anthropology in reaction to developments in cognitive psychology. In this development the human mind does not just acquire and memorize all information that is available and presented to it in a given cultural environment. On the contrary, the relationship between culture and mind is a two-way process: the mind not only acquires cultural information, but also shapes and transforms it. In light of this the anthropologist can now assume that underlying all cultural forms and diverse belief systems is a similar cognitive architecture shared by all humans (cf. Visala 2011, p. 6).

The *fourth* development that contributed to the emergence of CSR was the emergence of cultural Darwinism, i.e., theories that could now use conceptual resources from Darwinian evolution to actually model cultural evolution. CSR combines all four of these developments in explaining cross-culturally recurrent patterns in religious belief and behaviour. This also implies that religious ideas and practices are informed by our non-religious cognitive systems working in different domains. On this view there is not a single type of 'religious cognition' or 'religious module' in the human mind. Rather, the same general causes that explain other non-religious features of human cognition also explain religious cognition.

Some of the most important conclusions gleaned from these developments have consistently implied that the study of religion is explanatory rather than interpretative in nature. For many this has signalled a worrisome shared antipathy toward hermeneutical and sociological approaches. The work of Pascal Boyer (2001), in particular, has often been criticized for this (e.g. Visala 2011, pp. 17ff.). The idea that the study of religion should aim to *explain* rather than *understand* religion, is already, I believe, highly problematic from the points of view of philosophy and the philosophy of science. Importantly, though, Boyer does claim that the phenomenon of religion is not to be considered *sui generis*, as an exclusive category all on its own. In this sense religion is considered by CSR as a part of human culture and human nature, and thus as a natural phenomenon (cf. Visala 2011, p. 17), which, I believe, is in direct continuity with the claims and arguments of traditional evolutionary epistemology (cf. van Huyssteen 2006, pp. 75–110). Also of crucial importance here is that religious beliefs and behaviours are seen as by-products of the modular cognitive system that had an adaptive function in our ancestral environment. Rather than being adaptive, religion, on this view, is a by-product of systems selected for other functions (cf. Visala 2011, p. 41).

Visala and others have asked important questions about the scope, the boundaries, and limitations of CSR explanations of religion. At the same time, they have made it clear just how important the distinction between *adaptationist* and *non-adaptationist* accounts of the origin of religion has become. In his discussion of these controversial issues Jeffrey Schloss has concluded correctly that, while the proposal for the non-adaptationist status of religion has received much attention, the idea of a 'hypersensitive agency detection device' (HADD), has arguably received the most attention in discussions within and outside the field (2009, p. 17). Schloss' most important criticism against CSR can be stated as follows: *the real Darwinian question of what selective regime accounts for these dispositions is largely separate from and considerably less addressed than the empirical demonstrations of their existence and operation* (p. 17).

Jeffrey Schloss and Michael Murray's (2009) book on CSR is proving to be especially valuable, not only for the ongoing interdisciplinary discussion of the scope of CSR claims, but also for theologians who are interested in the evolutionary status of religion. Murray (2009), in his own essay on the scientific explanations of religion and the justification of religious belief, puts it forcefully: 'These models [that are now used to explain religion], if correct, show *not one thing more* than that we have certain mental tools (perhaps selected, perhaps 'spandrels') which under certain conditions give rise to the belief in the existence of entities which tend to rally religious commitments. But, pointing that out does nothing, all by itself, to tell us about whether those religious beliefs are *justified* or not... The mere fact that we have beliefs that spring from mental tools selected for by natural selection is, all by itself, totally irrelevant to the justification of beliefs that spring from them' (p. 169).

THE EVOLUTION OF THE HUMAN SELF

Before returning to more specific aspects of the current discussion of the evolution of religion, I want briefly to mention a number of intriguing contemporary proposals for some of the most defining traits of what makes us human. Such traits as these, I believe, deserve to be included in an embodied definition of human species specificity. They have profound implications for our understanding of both human nature and religion. As I hope to show, their focus on specifically bodily processes contrasts sharply with CSR's own emphasis upon natural cognitive tools such as HADD in explaining the evolution of religion.

In her fascinating studies, especially her acclaimed *Roots* series, Maxine Sheets-Johnstone goes to the evolutionary roots of human cognition, language, and communication by developing a very pronounced 'hermeneutics of the human body' (1990, 1994, 2008). Throughout all her work the emphasis is on the role of the human body in understanding meaning and mind. The most fundamental theme advocated in her work, however, is the existence of a biological disposition to use one's own body as a semantic template for the way we communicate with others. Sheets-Johnstone wants to go even further, however, and 'get back' to our hominid roots through the evolutionary continuum of the tactile-kinaesthetic body (cf. 1990, p. 280), so as to illuminate the evolution of sexuality, language, music, and morality.

Ultimately Sheets-Johnstone argues that the biological disposition to use one's own body as a semantic template to communicate with others is essentially about symbolic behaviour and conceptual origins (1990, p. 3), or what Steven Mithen (1996) would call the emergence of cognitive fluidity in the embodied human mind (pp. 70, 136). Her fascinating thesis is that human thinking is modelled on the body, and it is precisely the sensorily felt and sensorily feeling body that serves as the cognitive source of those human concepts that continuously shaped human thinking and evolution (Sheets-Johnstone 1990, pp. 5–6). It is against this background that Sheets-Johnstone argues that many contemporary answers to the origins of language fall short as they continue to ignore the reality that 'no language can be spoken for which the body is unprepared' (p. 135).

No wonder, then, that since the path from embodiment leads directly to the evolution of empathy and the moral sense, the fundamental embodied roots of morality have been of utmost importance in her most recent work. For Sheets-Johnstone empathy goes to the evolutionary heart of human personhood. Our ability to care, to trust, to empathize and have deep feelings for others is essential to the development of moral awareness and the realization of a fully resonant human being (2008, p. 193). As such it gives us access to the mental acts and processes of others, and through empathy we discover the feelings and values of others, what their convictions are, and precisely through this capacity for empathy we ultimately share what Husserl already described as an intersubjective world, that is, a *communally intelligible world* (p. 194). And of crucial importance, although often neglected or even forgotten, is that in empathy we basically make sense of each other in ways *outside* of language. (cf. Sheets-Johnstone 1999).

Here now, a deeply embodied sense of empathy clearly emerges: the core phenomenon we observe when we observe or reflect on our own empathetic actions is not simply a *body*, but a *moving body*. Movement is an *unfolding dynamic event* and we in fact have *perceptions* of movement as well as *feelings* of movement. Sheets-Johnstone accurately describes this as a *double spatial sense of movement*: we *perceive* our movement as a kinetic three-dimensional happening; and we *feel* the qualitative dynamics of our movement kinaesthetically (2008, p. 199). It is in this sense that empathy is through and through indicated somatically: it is indeed not merely a question of a body, but of a body in movement. Indeed, in empathy *we move in ways we are moved to move* (p. 207). In completing this argument, Sheets-Johnstone now calls for philosophical and evolutionary ways to understand how empathy is indeed a spontaneous outgrowth of affect attunement (p. 211). Empathy also has direct links to *Theory of Mind*, i.e. our intuitive knowing of the minds of others. For Sheets-Johnstone this is now enriched by redefining Theory of Mind as getting to know not just other minds, but *getting to know other moving bodies* (p. 213). In principle Sheets-Johnstone offers us a fascinating programme for how to engage any project in anthropology or ethics that opens up space in understanding for a particular form of interanimate meaning, namely *empathy, moral awareness*, and *a rationality of care* (p. 215). It is this rich re-visioning of a profoundly embodied notion of empathy, I believe, that will eventually become a crucial building block for understanding the evolution of religion.

EMBODIMENT IN COGNITIVE SCIENCE AND NEUROSCIENCE

In the quest for understanding the 'nature of human nature', philosophical work like Sheets-Johnstone's on empathy and morality has recently received strong support from some cognitive scientists and neuroscientists. In the field of cognitive science embodiment refers quite specifically to understanding the role of an agent's own body in its everyday, situated cognition. At the heart of this concept is the question of how our own moving bodies influence the ways in which we think and speak (Gibbs 2006, p. 1). But, what must a body be like for it to support cognition, language, empathy, and consciousness? This question becomes extra-important once we acknowledge that one of the traditional beliefs in the cognitive sciences is that intelligent behaviour, including the ability to think, perceive, and use language, need not arise from any specific bodily form (Gibbs 2006, p. 2). On this view, cognitive systems are best characterized in terms of their *functional states* (i.e., their logical and computational processes) without concern for how these states are physically realized. This traditional conception of mind and body has imposed serious limitations on the scholarly study of mental life in cognitive science (Gibbs 2006, p. 3). In fact, until recently, very few scholars have actually emphasized the importance of kinaesthetic action in theoretical accounts of how people learn, perceive, think, experience emotions and consciousness, and use language. In his book, Raymond Gibbs advances the idea that the traditional disembodied cognitive science view of mind is mistaken, because human cognition is shaped by embodied experience at the most fundamental level.

Gibbs' strategy for exploring the significance of embodiment is based on what he calls the '*embodiment premise*': people's subjective, felt experiences of their bodies in action provide part of the fundamental grounding for language and thought. Cognition is what occurs when the body engages the physical, cultural world, and must be studied in terms of the dynamical interactions between people and the environment.

EMBODIMENT IN NEUROSCIENCE

The works of Maxine Sheets-Johnstone and of Raymond W. Gibbs Jr., find significant support in German neuroscientist Thomas Fuchs' (2009) recent development of an *embodied cognitive neuroscience*. The embodied cognition perspective sees mind and brain as a biological system that is rooted in bodily experience and interaction with other individuals. Embodiment here refers to both the embedding of cognitive processes in brain circuitry and to the origin of these processes in an organism's sensory-motor experience. What emerges here, too, is a human body that is connected to its environment and to other embodied human beings. Fuchs thus strongly opposes any view that would see the human mind as somehow localized in, or caused by, or identifiable with, the human brain: this kind of short-circuit between mind and brain leads to a conceptual and methodological impasse, for it misses the essentially embodied, relational, and biographical character of the human mind.

For Fuchs, then, consciousness does not develop in an isolated brain, *but only in a living organism enmeshed in its environment* (2005b, p. 115). This could also be stated as follows: the human mind is not confined within the human head, but extends throughout the living body to the life world in which both mind and brain are embedded. The mind as such, therefore, includes the world beyond the membrane of the organism, especially the interpersonal world of self and other, which is also the world in which mind and brain are essentially formed (Fuchs 2009, p. 2). For Fuchs the brain is never the 'origin' of the mind, but rather acts as a *transformer* which translates the stimuli (single elements of a given situation) into wholes or *Gestalt* units, and finally into three cycles of embodiment: the affective self, the ecological self, and, ultimately, the intersubjective self. But in this context the brain is only an organ, and it is not just the brain, but the whole organism of the living person that has conscious access to the world.

Most importantly, and in direct support of Maxine Sheets-Johnstone's notion of embodied empathy, Fuchs talks about our bodies as being *transparent to the world*: human subjectivity is embedded in the world, with the body acting as its mediator. And consciousness, as the luminosity which reveals the world to a subject, is the direct result of this mediation. Miraculously, our bodies, as solid and material objects, are capable of a transformation that turns matter into mind and lets the world appear. In this way the body becomes transparent to the world and allows us to act in it (Fuchs 2005a, p. 95). For Fuchs this goes to the heart of emotional contagion and empathy, for we understand the gestures and facial expressions of others immediately: there is an implicit resonance with the expressions of others, while our own bodily and emotional reactions through emotional contagion show how the body works as a tacitly 'felt mirror' of the other. The discovery of 'mirror neurons' in the premotor cortex seems to provide the core neural mechanism of this sensorimotor integration. But observing the other's movements and gestures implies a transmission of intentions as well: we use the operative intentionality of our body as an instrument for understanding the other's intentions (p. 98).

EMPATHY AND ATTACHMENT

The important work on human personhood, and specifically embodied empathy, by scholars as diverse as Maxine Sheets-Johnstone, Raymond W. Gibbs Jr., and Thomas Fuchs, finds a particularly exciting enhancement in the work of psychologist of religion, Lee A. Kirkpatrick. Kirkpatrick broadens notions of empathy to include attachment theory. His work extends not only to the embeddedness of empathy in evolutionary history (as already became very clear in the work of Frans de Waal (2006)), but also to direct implications for religious belief, and particularly for the evolution of religion. In his book, Kirkpatrick (2005) addresses seminal questions such as, why has religion played such a strong role in all human cultures throughout history? Despite the remarkable diversity of forms of religious belief, why have certain common themes consistently emerged? In this work Kirkpatrick consistently places psychology of religion in a larger evolutionary context, and within this framework, attachment theory provides a powerful lens through which to reconceptualize and advance the embodied aspects of religious belief and behaviour.

Rejecting the notion that humans universally possess religion-specific instincts or adaptations, Kirkpatrick argues that religion should instead be seen as a by-product of numerous psychological mechanisms and systems that evolved for other functions, including empathy and attachment. Applying attachment theory to religion, Kirkpatrick identifies key parallels between early attachment relationships and adult romantic relationships, on the one hand, and an individual's perceived relationship with God, on the other hand. Seeing God as an attachment figure offers new ways of thinking about such core religious phenomena as images of God, prayer, religious development, and conversion. Evolutionary perspectives are now greatly influencing the ever-increasing popularity of attachment theory. In fact, for Kirkpatrick the emerging evolutionary perspective attempts something really new: *the tying together of attachment, love, caregiving, and mating into a larger, coherent framework* (cf. Kirkpatrick 2005, p. 51), which now includes a very distinct religious dimension. In fact, Kirkpatrick wants to argue that attachment processes are involved in many aspects of religious belief, and also directly in the evolution of religion itself.

Furthermore, and importantly for any discussion of the evolution of religious belief, this approach serves as a reminder that the human attachment system is only one of numerous evolved systems for regulating cognition, emotion, and behaviour in functionally distinct classes of relationships (cf. Kirkpatrick 2005, p. 74). Kirkpatrick is suggesting that the cognitive-emotional machinery of the attachment system provides a kind of deep structure or universal grammar for thinking about gods or other deities. This cognitive machinery is then employed readily in manipulating these ideas and drawing further inferences from them. Because the attachment system is species-universal, the influence of this same deep structure is evident in the beliefs about gods of many different cultures. At the same time, however, the parameters of the attachment system are set differently in different people by virtue of actual experience, giving rise to individual differences in some of the details of religious belief (Kirkpatrick 2005, p. 126). There is also increasing evidence that attachment, with a clear adaptive function in infancy and childhood, functions differently, for evolutionary reasons, in adulthood. In adulthood attachment is 'the tie that binds' certain relationships, especially romantic relationships and deep friendships, together (Kirkpatrick 2005, p. 201). In this sense, the attachment system—already in place for infancy and childhood—was adopted by natural selection as a suite of evolved mechanisms already well designed for the purposes of producing powerful emotional bonds, motivating commitment to a relationship, and maintaining proximity between two individuals. Natural selection, in effect, reassigned the system a new function in adulthood. Kirkpatrick sees this process of *exaptation*, in which an adaptation for one function is later co-opted and further evolved for use in solving a different adaptive problem, as indeed a common evolutionary process (2005, p. 201).

When Kirkpatrick finally turns to the evolution of religion he can now make explicit a general perspective that has been implicit throughout his work: religion is not itself an adaptation; humans do not possess, as part of our species-universal evolved psychological architecture, mechanisms designed by natural selection specifically for the purpose of generating religious belief or behaviour as a solution to any particular adaptive problem (2005, p. 238). Attachment theory and its embodied empathetic disposition as outlined here should be seen as part of a much broader model in which the attachment system represents just one of many domain-specific psychological mechanisms that have been co-opted in the service of religion and religious belief. Religion activates attachment processes but also many other psychological processes as well, and it is probably this *combination* that is responsible for its widespread success and staying power (Kirkpatrick 2005, p. 239). The path from genes to religious belief is, therefore, clearly a very long and circuitous one (p. 327).

What is interesting for an interdisciplinary theologian like me is that Kirkpatrick leaves open the possibility for positively interpreting the value and integrity of religious belief as he warns against the so-called *veridicality trap*: the common but patently false assumption that if certain beliefs can be understood and explained scientifically, then the beliefs themselves are by implication false. On the contrary, there is no reason why any scientific approach to understanding religion need assume that the beliefs under study are either ontologically true or false. To believe that any naturalistic explanation of the origins of a belief, neuro-psychological or otherwise, necessarily implies the belief is not true (because it has been 'explained') is a classic example of the *genetic fallacy* (Kirkpatrick 2005, p. 353).

In fact, the human brain/mind was designed according to the sole criterion of inclusive fitness, and is thus designed to be *adaptive*. It is decidedly *not* designed to be 'accurate' or 'correct' as judged by logical or other empirical standards. Often this leads to correct intuitions and inferences, and being 'correct' is indeed often adaptive. Once this is acknowledged, there is no *a priori* reason to believe that any particular kind of belief, whether religious or not, should be expected to be correct or incorrect. The mind is designed in such a way that, depending on any number of factors, it sometimes draws correct inferences and sometimes incorrect ones. In this sense an evolutionary psychology of religion should address the question of why and how people come to hold (or, come to reject) particular beliefs in which we are interested, irrespective of the question of whether or not they are true or false (cf. Kirkpatrick 2005, p. 354).

NEUROSCIENCE AND RELIGIOUS EXPERIENCE

A radically different view from Lee Kirkpatrick's non-adaptive account of the human propensity for religious awareness and religious belief, but one equally focused on human embodiment, is found in the work of neuroscientist Patrick McNamara. For McNamara it is exactly the deep religious propensities of the human mind that cannot be explained by naturalistic evolutionary accounts of human nature and behaviour only. An interesting and rather bold move in this direction was recently presented in his book *The Neuroscience of Religious Experience* (2009). As a neuroscientist McNamara wants to develop his own central conviction that religion is a defining mark of what it means to be human, as emblematic of its bearer as the web for the spider (p. ix). The special focus of McNamara's work, however, is the examination of religion through the eyes of the human self. Strikingly, in spite of the self's great dignity and worth, it is still treated by religions as divided, conflicted, and in need of salvation. McNamara argues that there is a considerable anatomical overlap between the brain sites implicated in religious experience and the brain sites implicated in the sense of 'self' and self-consciousness. This accounts for the crucial conclusion that religious practices often operate in such a way as to support the transformation of self so that it becomes more like an 'ideal self' whom the individual hopes to become (p. xi). In this sense religious practices directly contribute to the creation of a unified self-consciousness and to what McNamara calls an ideal 'executive self'. So, when religions are operating normally, they tend to create a healthy, unified, and integrated sense of self. Religions accomplish this feat by promoting a cognitive process that McNamara calls *decentering* (2009, p. 44).

In short, McNamara's bold claim is that religion is irrevocably a central part of the evolution of symbolic and religious behaviour, and of the construction of a centralized, 'executive' self. This implies that religion is not, as is often argued, an unfortunate by-product of more useful cognitive capacities of the human mind. On the contrary, this implies that religion is an adaptation, and, for McNamara, this is confirmed by the fact that the practice of religious rituals and belief in supernatural agents occur in virtually all human cultures (2009, p. 249). It is precisely religion's impact on the problems associated with the self and consciousness that could be seen as adaptive. From this perspective, the self and its default, divided, position should be taken into account when discussing the evolutionary history of religion (p. 253).

THE FAITH INSTINCT?

Patrick McNamara's neuroscientific argument for the adaptive status of religion has recently found interesting and converging support in the work of well-known *New York Times* science writer, Nicholas Wade. In his book, *The Faith Instinct: How Religion Evolved and Why it Endures* (2009), Wade has made the point that religions normally point to the realm of the supernatural, thus assuring people that they are not alone in the world. Many—believers and atheists—still find it difficult to understand religious behaviour from an evolutionary perspective: people of faith may not like the idea that the mind's receptivity to religion has been shaped by evolution; those who are hostile to religion often do not embrace the idea that religious behaviour evolved because it might have conferred essential benefits on ancient societies and their successors (cf. Wade 2009, p. 5). Wade, however, argues explicitly that an 'instinct for religious behavior' is indeed an evolved part of human nature. Because of the definite survival advantage conferred on people who practiced a religion, the behaviour—whether adaptive or non-adaptive—became written into our neural circuitry at least 50,000 years ago, and probably much earlier (pp. 5–6). Or, as Wade puts it, religion is a complex cultural behaviour built on top of a genetically shaped learning machinery (p. 5). People are born with the innate ability to learn the language and religion of their communities, and in both cases culture supplies the content of what is learned. This is also why languages and religions differ so widely from one society to the next, while remaining so similar in their basic form.

Against this background Wade's definition of religion emerges: religion is a system of emotionally binding beliefs and practices in which a society implicitly negotiates through prayer and sacrifice with supernatural agents, securing from them commands that compel members, through fear of divine punishment, to subordinate their interests to the common good (p. 15). As to the crucial and defining role of *morality* in religion, Wade thinks religion and morality share a common feature that reflects their origins as evolved behaviour: both are rooted in the emotions (p. 17), and both religious knowledge and moral intuitions appear in the mind as strong convictions, not as neutral facts. Wade argues that morality is older than religion, since we now know of its roots in primate behaviour. In this sense one could say that religious behaviour was engrafted on top of the moral sense in the human lineage alone (p. 17). Understanding how moral intuitions evolved thus makes it easier to see that religious behaviour also has an evolutionary origin. Frans de Waal's work on the building blocks of morality in primate behaviour, and its direct links to empathy and reciprocity already argues for this important fact. Clearly, hominids and later humans would have inherited these building blocks from their apelike ancestors and developed them into moral instincts. Evolutionary biology thus gives us a fascinating new explanation for morality: moral behaviour does not originate from 'outside' the human mind or even from conscious reasoning, sources often favoured by theologians and philosophers, but rather has been wired into the genetic circuitry of the mind by the process of evolution (cf. Wade 2009, p. 19).

EMBODIMENT AND THE EVOLUTION OF RELIGION

In Wesley J. Wildman's important book, *Science and Religious Anthropology* (2009), various strands from our conversation so far now flow together in an exciting and challenging way. The central argument of this work, not surprisingly, supports a naturalistic interpretation of the human being, but, quite specifically, the human being as *Homo religiosus*. Religion in this specific sense suffuses every aspect of human life, and Wildman makes it clear that our value commitments, our efforts at meaning construction, and our socially borne explorations of life possibilities, all reach far beyond the historically most prominent forms of religiosity. Wildman, in language reminiscent of Patrick McNamara, puts it well: at an axiological level, beneath the most overt beliefs and practices of both religious and non-religious people, we find *Homo religiosus*. From there on we can trace the impact of human religiosity in a more general sense on existential levels, on moral awareness, and on the social construction of reality (Wildman 2009, xvii). Wildman thus wants to present a religious anthropology by focusing on the embodied *Homo religiosus*, and by enlarging the scope of religious behaviours, beliefs, and experiences, to encompass everything relevant to human meaning and value.

Wildman also wants to ask about the *evolutionary status of religion*. He first considers two extreme views, namely, (i) that religion has no genetic component, or (ii) that religion is an adaptation rising on the back of one or two traits. Finally he develops his own, and in my view most plausible view, *that religion in evolutionary terms is a combination of side-effects of both adapted and non-adapted features of the human organism* (p. 37). Whatever one's views of theological truth claims, the evolutionary story of the origins of religion is directly relevant to assessing the meaning and value of religion, as well as religious claims about human beings (p. 42). In light of this, Wildman now asks, is religion an evolutionary adaptation, increasing fitness in and of itself and originating because of the adaptive functions of religious behaviours, beliefs, and experiences? Or, is religion a side-effect of a collection of adapted traits? Or is it possible that religion has no genetic component at all and finds its origin only in a long history of cultural expressions?

There is overwhelmingly strong evidence against the extreme view that no aspect of religion is genetically constrained or evolutionarily conditioned. Similarly, for Wildman it is clear that the evidence for religion as an adaptation so narrows the focus to one or two adapted 'religion traits' that only a fraction of the varied phenomena of religion are registered in the explanation. He argues, this time in language reminiscent of Lee Kirkpatrick, that it seems highly likely that the evolutionary explanation for the origin of the multifaceted reality of religious behaviours, beliefs, and experiences, must lie somewhere between these extremes. Religion, in other words, is evolutionarily conditioned, possibly in a few special respects by virtue of the adaptiveness of specifically religious traits, but in most respects by virtue of side-effects of traits adapted for some other, primarily and originally non-religious purpose (cf. Wildman 2009, p. 48).

In the end, Wildman states that the religionist or theologian gains little by mastering the intricate debate over adaptations versus exaptations versus spandrels, because little depends upon the details of how religion evolved *once it is granted that religion is in fact partly the product of evolutionarily processes*. The general fact that religion is partly the product of evolutionary processes indeed proves to be the most salient point for any religious anthropology (Wildman 2009, pp. 54–5). And at this point Wildman directly refers to Lee Kirkpatrick (2005) who, as we saw earlier, has argued that religion is a complex combination of side-effects that have a variety of adaptive functions. He states: 'Understanding religion in evolutionary terms predominantly as a combination of side-effects of both adapted and non-adapted features of the human organism, possibly with a few directly adapted features, is the hypothesis that I regard as possessing the most *prima facie* plausibility' (Wildman 2009, p. 56).

Crucially important from the perspective of this essay is Wildman's consistent and enduring focus on human embodiment. Because the human brain furnishes the cognitive, emotional, and motor capacities underlying our extraordinary range of religious behaviours, beliefs, and experiences, it is reasonable to expect the neurosciences to have at least as much transformative importance for religious anthropology as the evolutionary sciences (p. 87). And central to the current discussion in the cognitive science of religion is Wildman's argument that biases exist in the human cognitive system either because they have been selected in the evolutionary process for their survival benefits, or because they are side-effects of other traits selected for their usefulness (p. 94). This enables Wildman to focus on three domains in which a sharp awareness of human bodies and their functions is vital for understanding the human being as *Homo religiosus*: sociality, morality, and, finally, religious and spiritual experiences. In all three of these areas it is *bodies* that make religion, in the broadest sense of the word, not only possible, but also inevitable. Bodies shape the cognitive and emotional form that religion takes in individuals, and the social and moral practices that religions manifest in groups. Our bodies do not completely determine who we are, but they do constrain without determining, and as such they directly shape what we are and do, how we think and interpret, how we love, and how we construct religions (cf. Wildman 2009, p. 118). Thus our religious and spiritual experiences arise from a suite of bodily capacities with neurological and sensory roots that have vast existential and social impacts (p. 141).

Some scholars, of course, have argued that religion is an evolutionary phenomenon in a double sense, i.e., both on a biological and social/cultural level. Scholars like James McClenon (1997, pp. 345–54; 2002), and David Lewis-Williams (2002), for instance, have claimed from very different disciplinary backgrounds that altered states of consciousness, ecstatic religious experiences, and forms of shamanism are not only neurophysiologically grounded but represent the earliest forms of prehistoric religion. McClenon and Lewis-Williams have also provided possible scenarios for how shamanic rituals could have evolved by natural selection in the human ancestral environment, perhaps as early as 30,000 years ago. Lewis-Williams has also argued persuasively that beliefs in a supernatural realm persist worldwide. In addition it is exactly the persistence of religion into our modern, materialistic Western milieu that in fact points to the answer to the problem of the origin of religion: instead of religion being an answer to social and psychological needs, and in place of the supposed evolutionary stages of religion, he prefers to think of *origin-as-process* (2010, p. 137). I believe this plausible idea can be elaborated as follows: when reflecting on the evolutionary origins of religion, on the one hand, and the ongoing process of the evolution of religion, on the other hand, these two dimensions of the evolution of religion cannot be separated: the *reasons* why religion persists today are, in some fundamental ways, the same as those that explain why religion came into being in the first place.

COGNITIVE SCIENCE OF RELIGION AND FAITH?

In the new cognitive science of religion debate, a rather important shift in the 'evolution and faith' discussion seems to have developed during the past decade or two: instead of trying to show that evolution is true and then assert that Christianity must therefore be false, the new scientific study of religion attempts to use evolution to show that religions generally are by-products or 'accidents' of natural selection. On this view, as we have seen, evolution has endowed humans with particular mental facilities and social arrangements that prop up religious illusions. And the basic thesis is that if evolution can explain religious belief, then we have no need to appeal to the reality of God(s) or to account for those beliefs (Barrett 2009, p. 76).

It is interesting, however, that someone from within CSR's ranks is actually trying to show that religious faith (in this case the Christian faith) is compatible with the findings of cognitive science of religion. In an argument that should be important to theologians, Justin Barrett, rather than seeing cognitive and evolutionary explanations of religion as hostile to Christianity, sees much promise in the cognitive sciences to enrich our understanding of religion and religious belief. In his recent work, Barrett gives an account of why beliefs in superhuman agents (gods) is historically and cross-culturally natural and universal by appealing to the naturally occurring properties of human minds. For Barrett, cognitive science of religion need not be an enemy of Christianity, and could actually prove to be compatible with (orthodox) Christian theology (Barrett 2009, p. 77).

A fascinating fact, of course, is that one recurrent and distinctive mark of religions is that gods frequently become connected with moral concerns (Barrett 2009, p. 88). Why is that? The answer for Barrett is found in the fact that naïve biology and naïve physics constantly assist us in our chronic need to make meaning, and mental tools guide us to search the environment for causal explanation. When these explanations fail because of unfortunate or bad events like sickness and death, HADD, Theory of Mind, and other mental tools engage our intuitive desire for explanation. Most importantly, when agency is invoked, another mental tool kicks in, namely Intuitive Morality. Intuitive Morality generates non-reflective beliefs about what constitutes moral behaviour. In fact, from an early age, children appear intuitively to differentiate between moral codes and social conventions, and humans converge upon general rules of behaviour that typically frown on things like murder, adultery, theft, deception, treachery, and cowardice, especially as directed against one's own group (cf. Barrett 2009, p. 89). Clearly our god concepts will gain in reflective plausibility by working in concert with these non-reflective beliefs. This leads Barrett (2009) to three important conclusions:

(i) The naturalness of religious beliefs thesis is complete enough to begin exploring its implications for religious belief and its relationship with theological commitments.

(ii) As far as claims of the so-called new atheists are concerned: Barrett joins Schloss, Murray, and Goldberg, by stating that *explaining* how beliefs come about—no matter how complete the explanation—says nothing about whether a belief is true or *justified* (2009, p. 96). Belief in other minds and belief in gods are both highly intuitive consequences of cognitive architecture operating on ordinary inputs. Both are non-empirical, widespread beliefs. Neither is directly weakened by increasing scientific knowledge about how these beliefs come about any more than knowledge of the visual system makes us suspicious that the visual world is not really there (p. 96).

(iii) It is in this sense that the cognitive science of religion is not an automatic anti-religious tool or movement. In fact, one can consistently hold such an account of the evolutionary status of beliefs along with Christian theology. This does mean that thinkers with theological commitments have serious work to do (p. 97), but a Christian version of the cognitive science of religion remains possible (p. 98).

CONCLUSIONS

In this essay I have argued that the question of the evolution of religion and of religious behaviour can never be disentangled from the broader issue of the evolution of embodied human personhood. This implies that the evolution of distinctive traits and aspects of personhood like morality, sexuality, empathy, and the religious disposition played a defining role in the evolution of human communication and interpersonal attachment, and along with the evolution of complex symbolic behaviour, combine to give us important insights into the evolution of religion and religious behaviour.

I believe good arguments have been made for the fact that religion is not in itself adaptive. We humans do not possess, as part of our evolved neurological and psychological architecture, intuitive mechanisms designed by natural selection specifically for the purpose of generating religious beliefs or behaviour as solutions to particular adaptive problems. Accounts of empathy, Theory of Mind, attachment, altered states of consciousness, HADD, and the evolution of the moral sense/intuitive morality, show how these can all be seen as parts of a much broader model in which many domain-specific mechanisms have been co-opted in the service of religion and religious belief. Religion, we have seen, activates attachment processes, but it may also activate many other processes, including those concerned with altered states of consciousness and HADD. It may even be this combination of processes that is responsible for the widespread success and staying power of religious belief.

Religion, in other words, is evolutionarily conditioned, possibly in a few special respects by virtue of the adaptiveness of specifically religious traits, but in most respects by virtue of side-effects of traits adapted for other, primarily and originally non-religious purposes. Understanding religion in evolutionary terms predominantly as a combination of side-effects of both adapted and non-adapted features of the embodied human might be the most plausible hypothesis for beginning to understand the evolution of religion and religious behaviour. Thus religious and spiritual experiences arise from a suite of bodily capacities with neurological and sensory roots and vast existential and social impacts (cf. Wildman 2009, pp. 56–141). It is, therefore, significant in thinking through the theological implications of this debate that various scholars have argued (as we saw earlier) that cognitive approaches and adaptationist approaches can in fact be seen as complementary.

Reflecting on the scope and limitations of the CSR approach is important. Some scholars rightly feel that cognitive mechanisms (and thus CSR) can explain only population-level general tendencies and not particular cases of religious beliefs, practices, or traditions (cf. Visala 2011, p. 85). Another critical question might be whether CSR deals only with the most general features of religious thinking and behaviour and actually excludes everything else that embodied humans think and do, meaning that all distinctive features of religious traditions (as well as what religious people themselves identify as 'religious') actually seem to fall outside of CSR's scope (cf. Laidlaw 2007, 213; Visala 2011, p. 85). Most interesting for me, via evolutionary epistemology (cf. van Huyssteen 2006, pp. 45–109), is that finally Justin Barrett presents a certain type of defence of theistic beliefs by grounding them precisely in the reliability of our cognitive systems (cf. Visala 2011, p. 185). As theologians, both Niels Gregerson (2003) and I have both argued that the naturalness of religious ideas actually supports religious claims rather than undermines them: if religious beliefs are largely produced by normal human cognitive systems and if we generally trust these systems, then we should not suspect them in the case of religious beliefs (cf. van Huyssteen 2006, pp. 53–109; Visala 2011, p. 184).

As we have seen in the brief overview presented here, the evolution of empathy, attachment, and morality is, therefore, deeply intertwined with the evolution of the religious disposition, as well as with other distinctive traits of human beings. For religious believers this certainly means that, if one accepts that evolution has profoundly shaped the human body, why not accept that it has profoundly shaped our embodied minds too? What evolution has done is endow us with predispositions for asking ultimate questions. We are predisposed to search for meaning, and to learn and live the religion of a community, just as we are genetically predisposed to learn the language of our communities, and construct moral codes on the basis of our genetically predisposed moral sense. All of this goes far beyond CSR's narrow theoretical focus on particular cognitive modules such as HADD. And with religion, as with language and morality, it ultimately is culture, not genetics alone, which supplies the constructive content of what is learned, appropriated and believed.

REFERENCES

Barrett, J. 2009. Cognitive science, religion, and theology. In *The Believing Primate: Scientific, Philosophical, and Theological Reflections on the Origin of Religion.* Eds. J. Schloss and M. Murray. Oxford: Oxford University Press, 76–99.

Boyer, P. 2001. *Religion Explained: The Evolutionary Origins of Religious Thought.* New York: Basic Books.

De Waal, F. 2006. *Primates and Philosophers: How Morality Evolved.* Princeton: Princeton University Press.

Fuchs, T. 2005a. Corporealized and disembodied minds: A phenomenological view of the body in melancholia and schizophrenia. *Philosophy, Psychiatry and Psychology 12*(2): 95–107.

——. 2005b. Overcoming dualism. *Philosophy, Psychiatry and Psychology 12*(2): 115–17.

——. 2009. Embodied cognitive neuroscience and its consequences for psychiatry. *Poiesis & Praxis: International Journal of Technology Assessment and Ethics of Science* 6(3): 219–33.

Gibbs, W. R. Jr. 2006. *Embodiment in Cognitive Science.* Cambridge: Cambridge University Press.

Gregerson, N. 2003. The naturalness of religious imagination and the idea of revelation. *Ars Disputanti: The Online Journal for Philosophy of Religion 3*. Available online at <www.arsdisputandi.org>.

Kirkpatrick, L. A. 2005. *Attachment, Evolution, and the Psychology of Religion.* New York/London: The Guilford Press.

Laidlaw, J. 2007 A well-disposed anthropologist's problem with the 'Cognitive Science of Religion'. In *Religion, Anthropology and Cognitive Science.* Eds. H. Whitehouse and J. Laidlaw. Durham: North Carolina Academic Press, 211–46.

Lewis-Williams, D. 2002. *The Mind in the Cave: Consciousness and the Origins of Art.* New York: Thames and Hudson.

——. 2010. *Conceiving God: The Cognitive Origin and Evolution of Religion.* London: Thames and Hudson.

McClenon, J. 1997. Shamanic healing, human evolution, and the origin of religion. *Journal for the Scientific Study of Religion 36*(3).

——. 2002. *Wondrous Healing: Shamanism, Human Evolution, and the Origin of Religion.* De Kalb: Northern Illinois University Press.

McNamara, P. 2009. *The Neuroscience of Religious Experience.* Cambridge: Cambridge University Press.

Mithen, S. 1996. *The Prehistory of the Mind: A Search for the Origins of Art, Religion, and Science.* London: Thames and Hudson.

Murray, M. J. 2009. Scientific explanations of religion and the justification of religious belief. In *The Believing Primate: Scientific, Philosophical and Theological Reflections on the Origin of Religion.* Eds. J. P. Schloss and M. J. Murray. Oxford: Oxford University Press, 168–78.

Schloss, J. 2009. Introduction: Evolutionary theories of religion. Science unfettered or naturalism run wild. In *The Believing Primate: Scientific, Philosophical, and Theological Reflections on the Origin of Religion.* Eds. M. Murray and J. Schloss. Oxford: Oxford University Press, 1–25.

Schloss, J. and M. Murray. Eds. 2009. *The Believing Primate: Scientific, Philosophical, and Theological Reflections on the Origin of Religion.* Oxford: Oxford University Press.

Sheets-Johnstone, M.1990. *The Roots of Thinking.* Philadelphia: Temple University Press.

——. 1994. *The Roots of Power.* Chicago: Open Court.

——. 1999. Emotion and movement. A beginning empirical–phenomenological analysis of their relationship. *Journal of Consciousness Studies* 6(11–12): 259–77.

——. 2008. *The Roots of Morality.* University Park: The Pennsylvania State Press.

van Huyssteen, J. W. 1999. *The Shaping of Rationality.* Grand Rapids: Wm. B. Eerdmans.

——. 2006. *Alone in the World? Human Uniqueness in Science and Theology.* Grand Rapids: Wm. B. Eerdmans.

Visala, A. 2011. *Naturalism, Theism, and the Cognitive Study of Religion. Religion Explained?* Guilford: Aldershot Press.

Wade, N. 2009. *The Faith Instinct: How Religion Evolved and Why it Endures.* New York: The Penguin Press.

Wildman, W. J. 2009. *Science and Religious Anthropology.* Farnham: Ashgate Press.

9

Neither Friends nor Enemies: The Complex Relationship between Cognitive and Humanistic Accounts of Religious Belief

LÉON TURNER

RELATING THE DISCIPLINES

Introducing his book, *In Gods we Trust*, Scott Atran suggests that his task as an anthropologist, 'is to scientifically explain cultures and religions in terms of their material causes' (2002, p.10). Like so many others in recent years, Atran draws upon the theories and models of the cognitive science of religion (CSR) in pursuit of this goal. Naturalistic explanations of religion are nothing new, of course, but few can legitimately claim the levels of empirical and theoretical support that CSR claims, and even fewer have seen their influence spread so widely. CSR, it is safe to say, continues to cause quite a stir in religious studies, prompting much agonizing about the future of the field in the process. As Laidlaw and Whitehouse remark, 'The cognitive science of religion has claims on the professional interest of all anthropologists, and not only those with specific interests in religion, for it has implications for the most general questions in the discipline about the nature of human social life and of anthropological knowledge' (2007, p. 3). Consequently, in the twenty years since the publication of Thomas Lawson and Robert McCauley's seminal work, *Rethinking Religion: Connecting Cognition and Culture* (1990), a lot has been written about how CSR might be accommodated by some of the more traditional *modi operandi* of the humanities and social sciences—specifically, the exploration of religions' particular manifestations, from individual beliefs to the construction of particular institutional hierarchies, through the study of distinctive, historically and geographically contingent, sociocultural traditions, processes and institutions (cf. Laidlaw 2007; Slingerland 2008a, 2008b; Whitehouse 2007).

Many of those who believe these broadly sociohistorical approaches to the study of religion are compatible with CSR simply wish 'to bring these data-rich fields into commerce with explanatory resources of biology and cognitive science' (Schloss, Barrett, and Murray 2010, p. 625). Since CSR's origins are as much in social anthropology as contemporary cognitive science and evolutionary psychology, establishing its overall compatibility with sociohistorical approaches to religion in general, and with anthropology in particular, is still important for many of its supporters. Most importantly perhaps, by demonstrating its compatibility with these approaches CSR resists the common charge that it implicitly reduces the study of religion to the study of psychological structures and processes. After all, if cognitive accounts complement rather than directly challenge sociohistorical accounts then their focus upon those aspects of religion that have a clear cognitive component (such as belief in God, for example) is less likely to invite criticism for ignoring much of what the humanities and social sciences understand as 'religious'.

Compatibility, though, as social anthropologist James Laidlaw implies, can be interpreted in a number of different ways (2007, p. 216). Some, for example, wish to integrate the findings of experimental cognitive science and evolutionary theory into sociohistorical studies, letting each constrain the other in the production of a single overarching explanation of religion. These theorists presume that the humanities and social and cognitive sciences are ultimately studying the same thing, albeit from slightly different perspectives. Laidlaw himself offers an alternative perspective, suggesting that cognitive and humanistic approaches are compatible in the much more limited sense that they might peacefully co-exist, even if they cannot usefully interact with each other. The field of religious studies may be inherently pluralistic, he argues, but CSR works 'insofar as it does, by a methodological exclusion of much that we mean by religion' (p. 215). Its explanatory power is gained at the cost of the humanities' sensitivity to the historical contingency of particular religious traditions, which are shaped by the distinctively human capacity to exercise 'reason, imagination, and will' (p. 214). For Laidlaw, there may be intrinsic value to the cognitive theorists' work, but there can be no possibility of the coherent integration of cognitive science with the humanities. His position is quite different, therefore, both from those who would simply deny the validity of cognitive explanations of some aspects of religion, and those, including Scott Atran and Pascal Boyer, who see such explanations as crucial supplements to sociohistorical accounts. From such a perspective as Laidlaw's the most that could be gained from a conversation between CSR and the humanities is a marginally more detailed account of religion than the humanities are able to provide alone. Conflict between the disciplines is not seen as inevitable, but since there are such fundamental differences between the ways that cognitive scientists and scholars of the humanities go about their work, and because religion, being historically rather than analytically defined, is more properly an object of study for the humanities, significant mutual elucidation is unlikely (Laidlaw 2007, pp. 212–13).

Those convinced of the disciplines' compatibility typically argue that there is no conflict *in principle* between their distinctive approaches. Unfortunately, the absence of inevitable conflict as a result of their divergent epistemological and methodological commitments tells us little about the *de facto* compatibility between CSR and sociohistorical explanations of specific aspects of religion. It is a matter of some importance, since CSR is defined at least as much by what it tries to explain as how it tries to explain it. Curiously, relatively scant attention has been paid to the complexity of the relationship that exists between CSR and sociohistorical approaches to religion in particular areas, and regarding particular hypotheses. Rather, questions about methodological and epistemological compatibility tend to be tackled in the opening remarks or introductory chapters of works of CSR, often against the background of the interminable debate about the value of psychological and other 'scientific' explanations of sociocultural phenomena more generally. They are usually resolved long before more specific questions about specific aspects of religion are addressed.

I will argue below that, in practice, CSR often does court conflict with sociohistorical approaches to religion. Theoretical conflict, I will assume, exists wherever two different accounts of the same thing are mutually exclusive, or when one account directly or indirectly precludes the possibility of providing a compatible alternative, or implicitly or explicitly denies the value of such an alternative. In short, whereas some (perhaps most) of CSR's widely accepted cognitive explanations of various aspects of religion leave plenty of room for alternative, non-contradictory sociohistorical explanations, some clearly do not. This claim will be illustrated primarily through the analysis of Justin Barrett's extremely influential account of why people believe in gods. Barrett's work has been central to the development, enrichment, and dissemination of many of CSR's most important and best supported theories (see Barrett 2004, 2007, 2010). I will argue that some CSR theorists, including Barrett, seem comfortable with naturalistic causal explanations both of the origins of belief in gods, and of the pan-cultural recurrence of certain beliefs that explicitly and deliberately preclude the possibility of their explanation in terms of meaningful sociohistorical narratives, and therefore overtly conflict with a fundamental principle of (at least some) humanistic and social-scientific accounts. This in turn leads me to question the compatibility between sociohistorical accounts of religious belief and those CSR theories that suppose some beliefs in gods to be mere by-products of human cognitive evolution. Unfortunately, I will suggest, conflict between CSR and sociohistorical accounts of religion are likely to be most pronounced in precisely those areas where evolutionary psychology is presumed to make its most significant contributions.

CSR AND RELIGIOUS BELIEFS: THREE SORTS OF CLAIM

Underlying all aspects of CSR is the presumption that the activities of the cognitive system are strongly influenced by unconscious predispositions resulting from the normal functioning of (more or less) natural cognitive 'subsystems', 'tools', or 'devices'. Descriptions of these devices, their origins, and the ways in which they constrain the interpretation of reality in specific domains, vary between theorists, though most have tried to distance themselves from the so-called 'massive modularity thesis' embraced by early evolutionary psychologists (see Cosmides and Tooby 1992, 1994; Karmiloff-Smith 2000; Carruthers 2006). As Edward Slingerland (2008a) observes, opening the study of religion up to the methods of the natural sciences via these ideas about interpretative predispositions solves a number of perceived problems with traditional socio-historical approaches to religion. Most significantly, it breaks the study of religion out of the 'endless cycle of contingent discourses and representations of representations' that he believes characterizes the dominant social constructivist epistemology of the humanities and social sciences, and establishes 'a picture of the human person as an integrated body–mind system following the laws of nature' (pp. 378–9). Furthermore, it enables a connection to be made between our cognitive evolutionary history and the apparent ubiquity of at least some aspects of religion. This is made explicit, as we shall see, in accounts of how religion is constrained by natural cognitive resources—indeed, the concept of naturalness, in this sense, practically demands an evolutionary explanation.

Perhaps the most noteworthy of CSR's pursuits is its attempt to explain how a limited number of natural universal cognitive tools might be responsible for the automatic production of supernatural agent concepts in response to particular environmental triggers. When faced with the right sort of stimuli, which create explanatory voids that the individual perceptual system strives to fill, these tools simply encourage the interpretation of certain objects and events in terms of minded supernatural agents. The cognitive system is also presumed to constrain the spread or transmission of supernatural agent concepts between individuals, and ultimately to explain the apparent convergence of beliefs in supernatural agents amongst the world's religions. Central to this idea is the notion that, because our cognitive systems work in the particular ways they do, some concepts of supernatural agents are easier both to believe and to transmit than others. Both sorts of claim embrace the principle that, despite their apparent counterintuitiveness, supernatural agent concepts are processed 'offline' according to the same basic rules as regular agent concepts. Amongst other things, this is the basis of cognitive explanations of how complex belief systems are quickly constructed around concepts formed initially on the basis of limited concrete information, and why beliefs in gods and spirits appear to be culturally underdetermined.

There is a clear difference here between claims about the origins of beliefs in supernatural agents, claims about the ways in which supernatural agent concepts are represented and processed by individual minds after their acquisition, and claims about why individual minds are more likely to propagate some concepts rather than others. Together, they constitute the core principles of CSR's explanation of why belief in gods can be considered 'natural'. Summarizing this explanation, Geertz and Markússon write, 'religiosity (as belief in supernatural agents) is an emergent property arising from the *interplay* of normal cognitive mechanisms and the immediate natural and social environment (opaque causal processes → ideas → talking → spread of supernatural concepts)' (2010, p. 156). The remainder of this essay is devoted to the examination of each of these three sorts of claim in turn, with the overall aim of showing precisely how and where conflict might arise between CSR and sociohistorical approaches to religious belief. Claims about these three things, I will argue, are not all equally problematic from the perspectives of the humanities and social sciences. By showing precisely where intertheoretical conflict might arise, I mean both to demonstrate the complexity of the relationship that exists between CSR and sociohistorical approaches to religion, and to identify the particular issues likely to impede constructive interdisciplinary conversation.

REPRESENTING GODS

Let us consider, first, what is perhaps the least controversial aspect of CSR's explanation of the naturalness of belief in gods as far as the humanities and social sciences are concerned—the claim that individual representations of supernatural agents, such as gods or spirits, are constrained by the same cognitive processes that facilitate the efficient representation and processing of non-supernatural agent concepts. In this context, Barrett draws special attention to the notion of non-reflective beliefs: beliefs about the ways that people behave and think, which shape our everyday interpretations of the world around us, but of which we are seldom aware unless our attention is drawn explicitly to them. These are beliefs that 'come automatically, require no careful rumination, and seem to arise instantaneously and sometimes even "against better judgment"' (Barrett 2004, p. 2). In contrasting them sharply with reflective beliefs—'those we arrive at through conscious, deliberate contemplation or explicit instruction' (p. 2)—Barrett draws a distinction similar to that previously made by Dan Sperber between 'intuitive' and reflective beliefs (Sperber 1997). Barrett supposes that some non-reflective beliefs arise directly as cognitive responses to certain environmental triggers, some arise as a result of the 'automation' of extremely familiar or central reflective beliefs, and some are acquired culturally (albeit often completely unconsciously) through distinctive sociocultural traditions. But wherever they come from, non-reflective beliefs operate as 'cognitive defaults' in the absence of any more credible alternative reflective beliefs, or in the absence of conscious reflection upon those beliefs. 'Despite their mundane qualities', he suggests, 'non-reflective beliefs do a tremendous amount of work in filling out religious beliefs, motivating behaviors, and making the fancier theological notions possible' (2004, p. 10). Non-reflective beliefs about the ways non-supernatural agents work are easily and instinctively transferred to supernatural agent concepts, meaning that relatively few of the near-infinite number of beliefs people might theoretically come to hold about supernatural agents need to be taught directly. If religious beliefs in gods of various kinds can be considered a sub-category of belief in supernatural agents (and Barrett clearly thinks they can), we can conclude that 'All folk theology and religious practices gain structure and support from non-reflective beliefs' (p. 10).

The idea that people process information pertaining to supernatural agents in much the same way they process information about the other agents they perceive and interact with in their quotidian environments is a central pillar of CSR. It may only be of limited interest to humanists and social scientists, but it raises few if any challenges for them. They might perhaps argue about whether religion can be captured adequately in such individualistic, mechanistic psychological terms, but this is not strictly what information processing-based accounts of the representation of supernatural agent concepts intend. They do not strive to displace sociohistorical accounts of any particular dimension of religious belief, and they do not necessarily deny that sociocultural reality plays as important a role in the shaping of individual cognition as cognition plays in the shaping of representations of supernatural agents. They merely describe how a range of cognitive processes influence the representation of supernatural agent concepts, and how those concepts develop over time in relation to other beliefs and concepts. Hence, they are not in themselves explanations of what religious beliefs are, where they come from, or how they spread. They do not reduce religious belief to innate cognitive processes, and do not depend upon any particular claims about human cognitive evolution, other than the uncontroversial idea that the human brain's distinctive capacity to process information has been shaped by its distinctive evolutionary history. They are not even intended as comprehensive accounts of the ways in which religious concepts are structured, or the influences that might come to bear upon that structure. As partial accounts of a particular aspect of religion understood in the context of a particular kind of psychological discourse, they contain little to worry sociohistorical approaches at all.

Given their commitment to theoretical pluralism, and since Barrett's and others' experiments undoubtedly describe interesting aspects of beliefs in supernatural agents, including beliefs in gods, most of those in the humanities and social sciences would be content to see the findings of these studies added to the general repository of information about religion upon which subsequent theory might be developed (cf. Boyer 2010). At the very least, CSR's accounts of the ways in which normal cognitive processes constrain our thinking about religion can peacefully co-exist with sociohistorical accounts of religion, and they may even have some more constructive influence on some areas of religious studies. Although some might compare CSR's contribution, as Laidlaw does, to the sort of technical knowledge that materials science contributes to the study of art history (2007, p. 232), others will happily acknowledge the value of CSR's psychological explanation of, for example, how and why beliefs in supernatural agents are incorporated so seamlessly into everyday life. Certainly, there is unlikely to be any disagreement over the idea that our brains are able to represent and process information about supernatural and other agents in similar ways as a result of our unique evolutionary history. There is much greater scope for intertheoretical conflict, however, regarding some of the other roles that CSR typically attributes to the cognitive system in the production and transmission of beliefs in gods, and it is to these that we now turn.

THE ORIGINS OF BELIEFS IN GODS

In considering the contribution that CSR might make to our understanding of the *origins* of beliefs in gods, specifically, the particular processes by which non-reflective beliefs are produced, and questions about the extent to which these are influenced by sociohistorical factors take centre stage. As I have already observed, one of CSR's best developed theories concerns the idea that supernatural agent concepts are automatically produced by the coordinated functioning of various elements of a specialized cognitive 'tool-kit'. Two of the tools in this set have attracted the lion's share of empirical and theoretical attention: the so-called 'hypersensitive agency detection device' (HADD), the proper domain of which is the detection of agents exhibiting intentional goal-directed behaviour, and the 'theory of mind' (ToM), which is responsible for the attribution of mental states including desires and beliefs to those agents once they have been identified. There is a certain amount of disagreement between theorists over the precise roles they play (cf. Murray 2009, p.170), but as Barrett and many others have shown, the empirical evidence for their existence is substantial. Importantly, studies claiming to have identified their activity both in very young children and in well-educated adults appear to suggest that they are universal, maturationally natural, if not strictly innate features of our minds—an observation that points both to the evolutionary (survival) value of certain perceptual predispositions, and to the inescapability of certain ways of processing information (cf. Atran 2002; Bering 2011; Guthrie 1993, 2007).

Different theorists have emphasized the 'naturalness' of the specialized cognitive system to different degrees. At one end of the scale lies Scott Atran, who, in describing the evolved functions of the cognitive system, appeals to the ethological notion of the 'innate releasing mechanism', which describes the automatic production of animal behaviour in response to selected stimuli (2002, p. 60). Justin Barrett has always been less strident, rejecting outright the idea that the cognitive tool-kit is biologically 'hardwired': 'Mental tools and their processing biases arise primarily as a consequence of biological endowment plus essentially universal features of human environments' (2007, p.186). Elsewhere he states simply, 'That people the world over possess these mental tools does not necessarily mean that such tools are biologically "hardwired" into our brains or that their development is inevitable' (2004, p. 5). Boyer, like Barrett, draws a distinction between innateness and maturational naturalness, and suggests that all the latter implies about dispositions towards certain interpretations of the world is the naturalness of 'the capacity to form such understandings, given normal environments' (Boyer 2010, p. 378). McCauley (2000, 2011) employs the term 'maturational naturalness' in a similar way, and further distinguishes those capacities that arise naturally as a result of development in an amenable environment from those that become 'natural' as a result of extensive practice—those which evince 'practiced naturalness'. In this way, all emphasize the importance of environmental considerations in the development of supernatural agent concepts, and thereby circumvent the problem of genetic determinism whilst retaining the functional core of biological concepts of innateness and, crucially, preserving the theoretical link it potentially provides to evolutionary biological theory.

Psychological accounts of beliefs and behaviours presented in terms of the functioning of specific 'natural' cognitive tools, devices, or modules have traditionally been very unpopular amongst humanists and social scientists. But despite drawing so much attention, CSR's claims about the naturalness, or maturational naturalness, of certain cognitive resources don't always lead to intertheoretical incompatibility. As I have already suggested, there is no disagreement over the claim that the human capacity for religion has been strongly influenced by our evolutionary history, and no conflict regarding the claim that religious cognition is constrained by particular maturationally natural cognitive tools. Rather, the potential for conflict arises most clearly as a consequence of setting certain ideas about maturational naturalness in direct competition with the constructionist thesis that human beings' interactions with their environments are always mediated by *distinctive* social and cultural influences in several important ways. In some areas of CSR, this theoretical opposition succeeds in simply excluding the humanities and social sciences from discussions in which they believe they should be involved. This point can be illustrated by examining the role reserved for cultural influences upon belief (and cultural or sociohistorical modes of explanation) in Barrett's explanation of the cognitive origins of belief in gods.

According to Barrett, 'people believe in gods because gods gain tremendous support from the natural and ordinary operation of mental tools' (2007, p.186). More specifically, he claims that people are likely to invoke and believe supernatural explanations incorporating supernatural agents for various puzzling phenomena because supernatural agent concepts are rich in 'inferential potential'; they readily generate inferences and explanations about all sorts of things. They also fit nicely with the other sorts of intuitive non-reflective beliefs that the cognitive system spontaneously generates all the time. This increases their chances of becoming reflective beliefs, simply because the fewer non-reflective beliefs a particular reflective belief contradicts, the more likely it is to be believed: 'Reflective beliefs typically come from the cumulative weight of non-reflective beliefs converging on the same candidate belief' (p. 184).

As he elaborates his account of why people believe in gods, Barrett explains several different ways in which natural cognitive tools might encourage religious belief. Contrasting two of these will, I hope, help to identify with greater precision a source of potential conflict between CSR and sociohistorical approaches to religion. First, let us examine the natural support that Barrett believes certain cognitive tools lend to particular reflective beliefs. Citing the example of the part-human, part-goat entity known as 'Chivo man', belief in which is common amongst the members of a specific Californian community, Barrett explains how prior knowledge of the legend of this entity, in conjunction with the normal functioning of certain cognitive tools, might encourage local people to interpret ambiguous environmental events as further evidence of its existence (2004, pp. 27–8, 41–2). In fact, he even suggests that the possession of prior knowledge of Chivo man may actually heighten the sensitivity of HADD, increasing the chances that ambiguous stimuli will be interpreted in line with expectations. Here, the naturally occurring HADD and ToM tools are used to explain how the automatic generation of non-reflective beliefs in supernatural agents, which take a particular cultural form as a result of local traditions, can reinforce particular pre-existing reflective beliefs. The predisposition towards a certain sort of interpretation of the environment is given a psychological explanation, but the precise character of the subsequently generated beliefs are culturally or otherwise contextually determined, and require explanation in specifically sociohistorical terms—how the Chivo man legend began, how the legend is connected to other fragments of local folklore, and the enduring significance of the legend to the local community are all questions that defy purely psychological description.

Clearly, humanists' and social scientists' desire to interpret particular beliefs, such as Chivo man, as socially meaningful and historically contingent phenomena is not compromised by the psychological claim that these beliefs are actively encouraged by a proclivity to generate non-reflective beliefs about supernatural agents at the drop of a hat. As a psychological explanation of how people instinctively interpret their environments in a manner consonant with the supernatural beliefs of particular cultural traditions, it is no more problematic for them than psychological descriptions of any other everyday experience. It is different to, but totally compatible with a sociohistorical account of the development and persistence of the Chivo man myth.

Potentially more vexing, however, is the question of how maturationally natural cognitive tools like HADD are involved in the *generation* of novel supernatural agent concepts in the minds of individual observers. Geertz and Markússon, in critiquing the 'naturalness account' of religious belief they attribute to Barrett and others, capture its essence nicely in a simple thought experiment. They write, 'A group of infants find themselves on an uninhabited island where they grow up to form a social group in the absence of parental guidance. Will they grow up to be religious or not? The unnaturalness account would answer "No". There is no one to impart a religious tradition, no carrier of the "virus of the mind" called religion. The naturalness account would answer not an unequivocal "Yes", but a qualified "Most likely". The island's untouched, natural environment is filled with opaque causal processes, animal life, and such that are likely to stimulate various cognitive mechanisms, such as the ones that overextend animacy and agency. The generation of such thought processes leads to relevant ideas that in turn lead to talking, engendering in time the spread of ideas about unseen, powerful agents' (Geertz and Markússon 2010, p. 156). If CSR's supporters are really claiming that supernatural agent concepts—which we might think of as proto-god-concepts—are sometimes generated without any influence from the sociocultural environment at all, even if they are subsequently developed in distinctive sociocultural contexts, then this would seem to lead inexorably to conflict with at least some sociohistorical accounts of religion. Indeed, sociohistorical accounts of the origins of particular ideas about gods would not just be deemed superfluous in this instance, they would be rendered obsolete, as there would be nothing in such an explanation to be explained in sociocultural terms.

Here specifically, then, the question of intertheoretical compatibility depends upon whether and how CSR's supporters leave room for sociohistorical accounts to complement explanations of the psychological processes involved in the generation of proto-god-concepts. So how much room is actually left? Barrett (2007, p.192), for one, seems quite clear that some beliefs in and about gods (for example, the Christian belief in the trinitarian nature of God) are almost entirely the products of reflective rather than non-reflective beliefs. Yet he is also clear that, on occasion, certain ambiguous objects, such as 'wispy forms' that might bear passing resemblances to human figures, and certain ambiguous events, which lack immediately apparent causal explanations, do stimulate the production of novel supernatural agent concepts (2004, pp. 32–9). 'Without sufficient reflective defenses,' he writes, these non-reflective beliefs will eventually become reflective beliefs. How, then, is culture involved specifically in the origins of these concepts?

In his response to Geertz and Markússon's paper, Barrett clarifies his views on the ways in which culture impinges on the development of the cognitive system, without rejecting their characterization of the 'naturalness account' of the origins of religion. He writes, 'Cognitive accounts typically assume that common features of human environments play a critical role in helping to shape maturationally natural cognitive systems and fill in the content of religious beliefs and practices. Though maturationally natural cognition is fairly rigid, it typically requires a degree of tuning through environmental inputs. Once in place these cognitive systems do anchor cultural expression' (2010, p. 171). It seems, therefore, for Barrett at least, that universal sociocultural factors do have some role to play in the origins of *all* proto-god-concepts, even if he is less explicit than he could be about which environmental inputs 'tune' maturationally natural cognition, how 'tuning' works, or which features of sociocultural environments, relevant to the development of HADD and ToM, are truly universal.

Nevertheless, at least as far as the origins of some proto-god-concepts are concerned, the cultural realm is cast very much in a supporting role. Its influence is felt solely in the tuning of the cognitive system, nothing more. For Barrett (2010), culture, understood simply as 'a human social environment complete with artefacts, language, and other symbolic communication,' (p. 171) provides some of the basic resources necessary for HADD and other cognitive devices to do their jobs—by helping to delineate the range of opaque environmental stimuli to which HADD instinctively responds, for example. What is more, Barrett greatly restricts the sort of social and cultural processes involved here to those which are universal in human environments. The tuning of maturationally natural cognitive resources does not depend upon the particular ordering of social relations that characterize particular societies, particular rules governing the exchange of property or the division of labour, particular attitudes towards deceased ancestors, or any of the other historically and geographically contingent things humanists and social scientists might have in mind when they speak about 'society' and 'culture'. For Barrett, such distinctive cultural products (which presumably are developed reflectively, and transmitted mostly through special instruction) are just not necessary for the generation of supernatural agent concepts in the individual mind.

Barrett's notion of maturational naturalness recognizes the importance of exposure to empirical experience of the natural and social world, and does much to counter the charge, to which cognitive theorists and evolutionary psychologists are acutely sensitive, that cognitive accounts of religious beliefs are genetically deterministic. As Geertz and Markússon observe, the 'naturalness hypothesis' is rather a claim about the likelihood of religiosity in a given individual or particular community, rather than a claim about the necessity of religiosity (2010, p. 156). Tracks in the snow, for example, would not necessarily be attributed to the activity of unseen, and possibly even invisible, agents by someone who (quite implausibly) had never been exposed to footprints made by people or animals. But Barrett's account still ascribes the sociocultural world quite a different role in the generation of concepts of gods, to that it is usually ascribed by humanists and social scientists. It does not matter to Barrett what particular language is spoken, or what sort of rules govern particular narrative traditions, or even how people actually interact with each other in a particular community. Given sufficient exposure to human beings that do communicate somehow, appear to have thoughts and feelings, and causally interact with their environments, a person will naturally develop all the tools they need to generate (and subsequently pass on) novel supernatural agents of their own.

Conflict with the humanities and social sciences will not necessarily arise here as a result of Barrett's notion of the maturationally natural cognitive tool-kit, its ontogenetic or evolutionary history, or, once again, as a result of the idea that the tool-kit constrains the representation of beliefs. What is problematic, however, is CSR's occasional decoupling of particular, which is to say culturally derived, habits of mind from the interpretation of the environment in the moment of perception. It is this decoupling that characterizes CSR's naturalistic causal explanations of the origins of belief in gods, and distinguishes them from its accounts of how particular beliefs might be reinforced by natural cognitive tools, as in Barrett's discussion of Chivo man. In Geertz and Markússon's scenario, any particular concepts of gods produced by the island colony must emerge as a result of the need to discuss and explain the supernatural agents they cannot help but see evidence of all around them. There is no hint that the tendency to interpret their environment in this way can be influenced by anything culturally or socially distinctive about the colony itself.

Fundamental to sociohistorical modes of explanation is the notion that individual minds and their interpretations of their environments are always at least as much the products of historically and geographically contingent sociocultural processes as they are the products of individual cognitive and perceptual machinery. Humanists, social scientists and CSR theorists, it appears, often mean quite different things by 'context'. Although CSR recognizes the crucial importance of the local contexts of individual perceptual events to the final conceptual products of those events (as Schloss, Barrett, and Murray (2010, p. 625) observe, desert dwellers are unlikely to conjure up forest spirits), it seems primarily concerned with the ways these contexts constrain the availability of particular ambiguous natural stimuli necessary for supernatural agent concept generation (a physical environment devoid of trees prevents moving branches being a likely source of ambiguous stimuli, for example). And just because the actual generation of specific kinds of supernatural agent concepts depends upon particular, contextually determined, environmental stimuli, doesn't mean the rules governing their generation are not rooted firmly in more or less natural features of the individual cognitive system (cf. N. Barrett 2010), the relevant features of which develop in isolation from distinctive social or cultural influences. This is especially clear in Atran's appeal to the idea of the 'innate releasing mechanism'. CSR may recognize the importance of context in the cultural development of supernatural agent concepts once they have been produced, but this is orthogonal to its discussion of the origins of beliefs. Quite simply, for many CSR theorists, complex, richly culturally textured, and socially structured interpretations of particular environmental stimuli are simply not required for the genesis of proto-religious beliefs, either in the minds of contemporary individuals or their distant evolutionary ancestors. Ultimately, these theorists obviate the need for, and perhaps even the possibility of, any kind of sociohistorical description of how or why particular concepts of gods (or other supernatural agents) originated. Intertheoretical conflict does not arise simply because CSR generally favours a concept of mind that is naturally furnished with a particular set of cognitive tools, but specifically because of how that concept of mind is integrated into a causal naturalistic explanation of the ultimate origins of religious belief.

Many will not be surprised, of course, to find fundamental disagreement with sociohistorical approaches to religion in precisely that area where CSR is presumed to make its most startling and controversial claims. Some will even be surprised to discover there is so much compatibility in other areas. But the nature of the disagreement suggests certain other difficulties for those who wish to foster the ties between CSR and other branches of religious studies, whilst remaining committed to CSR's explanatory framework. These concern the bold (almost subject-defining) idea that religious beliefs are evolutionary by-products. There is, as I have already suggested, no need to worry about conflict between sociohistorical approaches to religion and CSR's accounts either of the natural predisposition towards interpreting the world in agentic terms, or its description of how maturationally natural cognitive resources constrain and reinforce certain aspects of beliefs in gods. Nor is there likely to be any serious objection to the claims that HADD's hair-trigger and the ToM device are evolutionarily adaptive, or that some of the interpretative anomalies they sometimes precipitate are mere evolutionary by-products. None of these notions obviate sociohistorical accounts of how and why people come to believe in gods.

But accounting for the origins of religious beliefs in gods and accounting for the evolutionary origins of the natural tendency to interpret certain ambiguous objects and events in agentic terms (which may support or reinforce belief in gods) are quite different enterprises. A causal explanation of the former in terms of the latter is potentially problematic for the humanities and social sciences if distinctive, non-universal sociocultural factors are simultaneously denied a role in the actual production of novel proto-god-concepts. But this is precisely what is implied by the claim that some concepts of gods originate as automatic and culturally unmediated responses to environmental stimuli, and are only subsequently shaped by distinctive social and cultural forces into their particular forms. Whether CSR's aim is to explain the generation of novel beliefs in new gods in the contemporary world, or to the emergence of the very first beliefs in gods in human prehistory, the humanities are simply excluded from the discussion.

Evolutionary theory in general, and evolutionary psychology specifically might have much to contribute to the study of religion in ways that do not conflict either with humanistic modes of explanation or particular humanistic explanations of particular phenomena. These might include accounts of how beliefs in gods are reinforced by cognitive mechanisms that may have enhanced survival prospects and reproductive success in various ways, and accounts of how cognition might itself have influenced the evolution of religion (cf. Whitehouse 2008). But one of its most significant contributions to our understanding of where beliefs in god come from, both in contemporary and prehistoric minds, is difficult to square with sociohistorical approaches to religion. Whether the cognitive resources responsible for the generation of proto-god-concepts are maturationally natural or innate in a stricter sense, makes little difference here. Neither concept ascribes the meaningful role to distinctive sociocultural phenomena that characterizes humanistic and social scientific explanations. Even the very term 'by-product' seems to isolate the origins of belief from sociohistorical explanation, in conveying the impression, as Laidlaw (2007, p. 229) observes, that religious beliefs begin as unconscious cognitive processing errors, not as the complex sociocultural products of historically contingent social and cultural forces.

If CSR wishes to avoid conflict with the humanities and social sciences (and it is certainly not the case that all CSR's supporters do), then it seems this particular evolutionary explanation of the origins of belief in gods must be abandoned. But to do so would dramatically change the explanatory scope of CSR. The very idea of the naturalness of religion, as Barrett and others understand it, depends upon evolutionary psychological ideas. Many of CSR's core theses, of course, need make no reference to the naturalness of religion at all—theories of how the cognitive system reinforces belief in gods need make no strong claims about the naturalness of that system. But a CSR stripped of its evolutionary account of the origins of belief in gods seems unlikely (much to Slingerland's undoubted disappointment) to finally break the humanities' 'endless cycle of contingent discourses and representations of representations'.

THE CROSS-CULTURAL CONVERGENCE OF RELIGIOUS BELIEFS

The sources of potential conflict between CSR and sociohistorical approaches to religion, however, are not limited to CSR's naturalistic causal explanation of the origins of belief in gods. CSR's attempt to explain perceived regularities amongst the world's religions, via an account of the constraints that the cognitive system places upon the spread of religious beliefs amongst populations, also raises certain difficulties. Explaining the transmission of beliefs is, of course, a crucial aspect of CSR's overall project since, as Barrett argues, 'If only one person has a belief in a particular superhuman agent, that is not a religious belief but an idiosyncratic one or an oddity. But when a group of people share the same beliefs in superhuman agents, then we have religious belief' (2007, pp. 186–7). In explaining the transmission of supernatural agent concepts from one individual to another, then, he is also explaining the production of religion itself. Since one of Barrett's aims is to provide a naturalistic explanation of the pancultural recurrence of certain features of religion, including, specifically, belief in supernatural agents of one sort or another, his account of belief transmission focuses upon the explanation of why certain beliefs are so successful. He recognizes that all *particular* beliefs in gods are socioculturally acquired, but he is especially concerned with explaining certain types of religious beliefs' differential success relative to the theoretically near-infinite range of all possible beliefs.

Barrett (2004) provides a host of reasons for the differential success of particular beliefs, including their reinforcement through religious 'actions' (such as ritual ceremonies) and their relationship to 'intuitive morality', but there is one factor he deems particularly important. For Barrett, successful beliefs are those that conform closely to the intuitive assumptions which he calls non-reflective beliefs, and which Atran (though he diverges slightly from Barrett's account in other ways) calls 'habits of mind'. Beliefs in supernatural agents, it is argued, enhance their inferential potential and memorability, and so facilitate their smooth transmission, through their effective recruitment of natural cognitive tools. Central to this thesis is the idea that supernatural agent concepts, simply by virtue of their so-called 'minimal counterintuitiveness' (they conform to the vast majority of intuitive beliefs about the world, but violate a small number of them), are especially 'attention-grabbing'. The marked improvement in memorability and ease of transmission enjoyed by minimally counterintuitive beliefs, coupled with a natural predisposition towards interpreting our environments in terms of supernatural agents that are rich in inferential potential explains not just the ubiquity of beliefs in gods, but also why beliefs in intentional minded agents, rather than anything more outlandish, are so prevalent. Effectively, the cognitive system acts as a filter, preventing those beliefs that do not conform to certain standards of minimal counterintuitiveness or inferential richness from entering circulation. The religious beliefs we actually see in the world today, then, exhibit certain recurrent features simply because those features are, and always have been, essential to their very survival. Whereas it might be possible for individuals to develop other (possibly more outlandish) sorts of beliefs reflectively, they are less likely actually to become religious beliefs, in the sense of being widely distributed, because it is so hard for them to pass from mind to mind.

What does all this mean for the compatibility of cognitive and sociohistorical accounts of religious beliefs? As a psychological description of cognitive constraints upon the transmission of religious beliefs, much of Barrett's account, once again, does not raise problems, since it does not preclude the possibility of alternative sociohistorically grounded explanations. These may be complementary to, or contrast sharply with, Barrett's account, but they are not excluded by it. However, the naturalistic explanation of cross-cultural religious regularities is a much stronger ambition than the mere demonstration of cognitive science's potential contribution to our overall understanding of the historical and contemporary prevalence of belief in gods. For some CSR theorists, including Atran, the aim is to explain these regularities in terms of a *universal evolved* cognitive architecture. These explanations find sociohistorical explanations much more difficult to accommodate.

Although Barrett accepts that some non-reflective beliefs might be culturally acquired, and he enthusiastically embraces the idea that particular concepts are culturally scaffolded by non-natural cultural artefacts, like most other CSR theorists, he is most interested in those non-reflective beliefs at the maturationally natural end of the nature-nurture continuum. After all, he claims, religion's recurrent features cannot be explained in terms of non-reflective beliefs that are themselves historically contingent and idiosyncratic: 'Recurrent cultural expression seems to require recurrent causes such as undergirding cognitive systems or environmental regularities' (2010, p. 171). From the perspective of Barrett's brand of CSR, then, particular cultural traditions have no role to play in determining which ideas are minimally counterintuitive and rich in inferential potential. They seem to play little or no part in the cognitive system's filtering activities. As Barrett observes, 'whether or not a concept is counterintuitive in this technical sense is largely or entirely independent of cultural context' (2007, p. 188).

Since the cognitive system's ability to function as a conceptual filter is assumed to be independent of cultural context, the possibility of a meaningful compatible account of the recurrence of certain features of religious beliefs in terms of particular, which is to say contingent, non-universal sociohistorical, factors, appears explicitly to be ruled out. Even if it is not denied in principle that certain cultural factors (certain culturally distinctive non-reflective beliefs, perhaps) might help determine what is minimally counterintuitive about any given concept at any given moment, these factors are still deemed superfluous to CSR's explanation of the universal features of religious belief systems. Whatever the humanities and social sciences might have to say about the differential success of actual beliefs, relative to other possible but unrealized sorts of belief, is just irrelevant to this sort of explanation. Once again, they are simply excluded from a conversation in which they would surely believe they could participate (even if their primary interests typically concern the particularities of distinct religious traditions rather than the similarities between them). From Barrett's perspective, had the whole of human cultural history been completely different, all extant religious beliefs, whatever specific form they took, would still share a common set of features.

This is still not to say that CSR denies the meaningfulness of religious beliefs, their embeddedness in broader cultural traditions, or the value of sociohistorical accounts of how historically contingent factors, both ecological and sociocultural, affect the spread of religion. Indeed, some of those who are largely sympathetic towards CSR have attempted to synthesize universal cognitive accounts of religious beliefs with an understanding of the sociohistorical particularity of religion in a broader sense (see, for example, Whitehouse 2004). CSR's accounts of the cognitive constraints upon religious transmission will not necessarily conflict with sociohistorical explanations of why one particular religion incorporating a particular minimally counterintuitive concept of god succeeded while another incorporating a different concept of god failed. But a causal explanation of the pancultural convergence of aspects of religious beliefs, which is presented exclusively in terms of maturationally natural and universal structures of mind, simply denies the relevance, and perhaps even the validity, of sociohistorical explanations. Once more, then, intertheoretical incompatibility between CSR, the humanities and the social sciences arises from the former's occasional predilection for causal naturalistic explanations of certain phenomena, not its model of mind, the narrowness of its concept of religion or its general ideas about maturational naturalness.

In light of these quite serious differences of approach and opinion, let us return again to the question of the extent to which evolutionary psychology can contribute to a cognitive science of religion that values its compatibility with the humanities and social sciences. Evolutionary psychology's distinctive contribution to this area is its grounding of universal constraints upon religious transmission in the natural evolved propensities of the cognitive system. This support's CSR's causal naturalistic explanation of the differential success of certain kinds of religious belief, and so the universal recurrence of certain sorts of supernatural agent concept. But if CSR is to avoid conflict with those disciplines that favour sociohistorical modes of explanation, then it must stop short of making these very strong claims. And if it does that, then evolutionary psychology's contribution to this area of CSR is greatly restricted. Evolutionary psychological accounts of the origins of the capacity to generate supernatural agent concepts and process them alongside other sorts of agent concepts, for example, are not uninteresting. Nor are evolutionary psychological accounts of the origins of those features of the cognitive system that enable some religious beliefs to pass from mind to mind so efficiently. But they add little to an account of religious transmission that takes the existence of such capacities for granted. If people do have these capacities, then, like any other kind of cognitive capacity, they must have been shaped to a certain extent by evolutionary pressures. The specific details of how and why they evolved contributes nothing more than superfluous detail to our understanding of why religions take the form they do. Severing the causal connection between the convergence of religious beliefs and evolved cognitive architecture, it appears, may reduce tensions between ECSR and sociocultural approaches to religion, but it leaves evolutionary psychology without much purpose at all in this area.

EVOLUTION, CONFLICT, AND COMPATIBILITY

To summarize, I have argued that the peaceful co-existence of CSR with humanistic and social scientific accounts of religion depends upon each leaving space for the other in their explanations of various aspects of religion. In practice, this does not always occur. As far as the origins of beliefs in gods are concerned, CSR theorists commonly embrace causal naturalistic explanations which simply preclude alternative explanations in terms of distinctive, historically contingent social and cultural features of the particular societies in which individuals develop. Similarly, CSR's attempt to explain certain perceived regularities between belief systems all over the world and throughout history in terms of the evolved cognitive capacities that enable the efficient transmission of minimally counterintuitive beliefs, leaves no space for sociohistorical accounts to play a complementary role. I have also suggested that conflict between ECSR and the humanities and social sciences arises in precisely those areas where evolutionary psychology is assumed to make some of its most significant contributions to the naturalistic explanation of religion.

And yet it is also clear that many core claims of CSR, including those relating to how beliefs in gods draw support from natural cognitive capacities and processes, raise no problems at all for the humanities and social sciences. The extent to which these may be usefully combined with sociohistorical accounts in enhancing our understanding of religion must be assessed on a case by case basis, but this essay has not identified any insurmountable epistemological or methodological reasons why they cannot, at least, peacefully co-exist. Nor is the relationship between the specifically evolutionary cognitive science of religion, the humanities and the social sciences necessarily one of outright conflict as far as the study of religious beliefs in gods is concerned. Explanations neither of the historical development of the particular sort of cognitive system that forms the foundation of much contemporary cognitive science, nor the ways in which that system both constrains the ways people think about gods, and even reinforces belief in them, need exclude sociohistorical explanations from the discussion. Many humanists and social scientists will undoubtedly object to them for many different reasons, but there is no need for them in principle to reject the idea that the mind really is possessed of the sorts of maturationally natural capacities that the cognitivists describe, nor the possibility that those capacities might have served important evolutionary functions that are unrelated to the production of religion. In short, it seems quite clear, if it was ever in doubt, that both CSR and ECSR enjoy complex relationships with the humanities and social sciences regarding the question of why people believe in gods.

What are the implications for the future of ECSR? I have not suggested, here, that ECSR's supporters should necessarily renounce the claims that humanists and social scientists are likely to object to; they are obviously not all obliged to get along! The extent to which CSR more generally actually depends upon evolutionary psychological ideas continues to be debated, but, if CSR continues to emphasize the natural foundations of religion in a way that excludes sociocultural explanations at key points, then its broader relationship with the humanities and social sciences will be severely, perhaps terminally, compromised. Given the importance that many working in the field ascribe to the maintenance of good relations, this may be too high a price for them to pay, but the alternative is perhaps not much more attractive. After all, it would seem to contradict many peoples' ideas of what CSR is, in practice, all about.

REFERENCES

Atran, S. 2002. *In Gods We Trust: The Evolutionary Landscape of Religion*. Oxford: Oxford University Press.

Barrett, J. 2004. *Why Would Anyone Believe in God?* Lanham, MD: Alta Mira Press.

——. 2007. Gods. In *Religion, Anthropology and Cognitive Science*. Eds. H. Whitehouse and J. Laidlaw. Durham, NC: Carolina Press, 179–207.

——. 2010. The relative unnaturalness of atheism: On why Geertz and Markússon are both right and wrong. *Religion 40*: 169–72.
Barrett, N. F. 2010. Toward an alternative evolutionary theory of religion: Looking past computational evolutionary psychology to a wider field of possibilities. *Journal of the American Academy of Religion 78*(3): 583–621.
Bering, J. 2011. *The God Instinct: The Psychology of Souls, Destiny, and the Meaning of Life*. London: Nicholas Brealey.
Boyer, P. 2010. Why evolved cognition matters to understanding cultural cognitive variations. *Interdisciplinary Science Reviews 35*(3–4): 376–86.
Carruthers, P. 2006. *The Architecture of the Mind: Massive Modularity and the Flexibility of Thought: Massive Modularity and the Flexibility of Thought*. Oxford: Clarendon Press.
Cosmides, L. and J. Tooby. 1992. Cognitive adaptations for social exchange. In *The Adapted Mind*. Eds. J. H. Barkow, L. Cosmides, and J. Tooby. New York: Oxford University Press, 163–228.
——. 1994. Origins of domain specificity: the evolution of functional organization. In *Mapping the Mind*. Eds. L. A. Hirschfield and S. A. Gelman. Cambridge: Cambridge University Press, 85–116.
Geertz, A. W. and G. I. Markússon. 2010. Religion is natural, atheism is not: On why everybody is both right and wrong. *Religion 40*: 152–65.
Guthrie, S. E. 1993. *Faces in the Clouds: A New Theory of Religion*. New York: Oxford University Press.
——. 2007. Anthropology and anthropomorphism in religion. In *Religion, Anthropology and Cognitive Science*. Eds. H. Whitehouse and J. Laidlaw. Durham, NC: Carolina Press, 37–62.
Karmiloff-Smith, A. 2000. Why babies' brains are not Swiss army knives. In *Alas Poor Darwin: Arguments against Evolutionary Psychology*. Eds. H. Rose and S. Rose. London: Jonathan Cape, 144–56.
Laidlaw, J. 2007. A well-disposed social anthropologist's problems with the 'Cognitive Science of Religion'. In *Religion, Anthropology and Cognitive Science*. Eds. H. Whitehouse and J. Laidlaw. Durham, NC: Carolina Press, 211–46.
Lawson, E. T. and R. N. McCauley. 1990. *Rethinking Religion: Connecting Cognition and Culture*. Cambridge: Cambridge University Press.
McCauley, R. N. 2000. The naturalness of religion and the unnaturalness of science. In *Explanation and Cognition*. Eds. F.C. Keil and R. Wilson. Cambridge, MA: MIT Press, 61–85.
——. 2011. *Why Religion is Natural and Science is Not*. New York: Oxford University Press.
Murray, M. J. 2009. Scientific explanations of religion and the justification of religious belief. In *The Believing Primate: Scientific, Philosophical and Theological Reflections on the Origin of Religion*. Eds. J. P. Schloss and M. J. Murray. Oxford: Oxford University Press, 168–78.
Schloss, J. P., J. L. Barrett, and M. J. Murray. 2010. Looking past vs. overlooking cognitive-evolutionary accounts of religion: A response to Nathaniel Barrett. *Journal of the American Academy of Religion 78*(3): 622–8.
Slingerland, E. 2008a. Who's afraid of reductionism? The study of religion in the age of cognitive science. *Journal of the American Academy of Religion 76*(2): 375–411.

——. 2008b. *What Science Offers the Humanities: Integrating Body and Culture.* Cambridge: Cambridge University Press.

Sperber, D. 1997. Intuitive and reflective beliefs. *Mind and Language 12*(1): 67–83.

Whitehouse, H. 2004. *Modes of Religiosity: A Cognitive Theory of Religious Transmission.* Oxford: AltaMira Press.

——. 2008. Cognitive evolution and religion: Cognition and religious evolution. In *The Evolution of Religion: Studies, Theories, and Critiques.* Eds. J. Bulbulia, R. Sosis, E. Harris, R. Genet, C. Genet, and K. Wyman. Santa Margarita, CA: Collins Foundation Press, 31–41.

Whitehouse, H. 2007. Towards an integration of ethnography, History and the cognitive science of religion. In *Religion, Anthropology and Cognitive Science.* Eds. H. Whitehouse and J. Laidlaw. Durham, NC: Carolina Press, 247–80.

Whitehouse, H. and J. Laidlaw. Eds. 2007. *Religion, Anthropology and Cognitive Science.* Durham, NC: Carolina Press.

10

The Cognitive Science of Religion from an Anthropological Perspective

TIMOTHY JENKINS

My aim is to consider cognitive scientific discussions of religion from an anthropological perspective. The chapter falls into two parts. In the first, I shall outline the positions of the two disciplinary conversation partners, adopting broadly a historical—though highly selective—approach: first, taking Durkheim's defence of the independence of sociology (Durkheim 1982), for this defence lies behind anthropological attitudes to psychology, and second, offering an account, drawn from Descombes (2001), of the development of the particular characteristics of the cognitivist approach. In the second part, which constitutes the centrepiece of the argument, I shall look at a particular example of how each discipline approaches a single topic, making a comparison between two explanations of witchcraft, the one—by Evans-Pritchard (1937)—anthropological, the other—by Cohen (2007)—drawing on cognitive science. I will conclude with some remarks on a possible way of envisaging cooperation between the two disciplines.

PART 1: TWO DISCIPLINES IN CONVERSATION

Durkheim's Defence of the Independence of Sociology

From the last third of the nineteenth century, sociology and psychology have maintained a sibling rivalry. Psychology is the older discipline; sociology however claims to explain a set of human behaviours independently of psychological theories; and since this initial claim, there have been a series of counter-claims by psychology to explain the behaviours reputedly set apart, and a corresponding series of sociological or anthropological defences. The classic statement of the independence of sociological facts from those explained by psychology and biology is made in Durkheim's *The Rules of Sociological Method* ([1895] 1982).

Durkheim first concedes that psychology since the 1860s has made an advance that sociology must imitate, in that it treats psychological phenomena as natural phenomena. By this he means that psychological facts are not considered from the perspective of the individual consciousness which experiences them, but as states of consciousness to be studied externally (Durkheim 1982, p. 71). Sociology ought the more readily to make this advance, to treat social facts as natural facts, because while it demands a certain effort to see internal mental states as such (it appears to do violence to their nature), social facts readily appear as things, things such as laws, statistics, monuments, records, and fashions. 'By their very nature', Durkheim writes, 'social facts tend to form outside the consciousness of individuals, since they dominate them' (p. 72). Durkheim's naturalism is a much discussed point; we shall need to see what value he gives to it.

Durkheim offers a critique of a series of writers who have tried to create accounts of human society on the basis of individual psychology. They suppose that individuals have needs and ideas which in turn determine the formation of societies, and that these needs and ideas include the notion of progress, or a teleological component. In this way, individual consciousness is the source of all social evolution, and psychological laws determine sociological laws. Human nature in this account creates collective life: in the case of simpler societies, you can read off the characteristics of social life from human nature directly, by deduction; in the case of more complex, progressive societies, you do so by inference after observation (p. 125). Beyond a general account of the progress of civilization, Durkheim adds, many theories of institutions work along similar lines, seeking to derive features of the family, marriage, the law, economics, ethics, and religion from individual feelings, advantages, reactions, and desires (p. 127). In this perspective, the cognitive science of religion may be seen as the latest in a long line of such theories, though the nature of the teleological tendency, the driver of the processes of evolution, is much debated, rather than being assumed.

Yet once such writers have conceded the need for observation in complex societies, they have accepted that human nature experiences alien constraints in the effects of the external environment. For once you cannot directly read off social features from natural ones, and have to have recourse to observation and inference, you are concerned with features that do not directly derive from human nature; instead the actors experience the authority of the social. The matter of authority or obligation is the distinctive feature of the social fact in Durkheim's argument, which he calls the pressure that such facts exert over the individual (p. 128).

The fact of obligation points to this issue: that the whole is not simply the juxtaposition of its parts, but instead an association which displays new properties, new properties which are the source of innovations, of things you must do, or ought to do, or can do, and so forth. By this principle, Durkheim writes, 'society is not the mere sum of individuals, but the system formed by their association (which) represents a specific reality which has its own characteristics' (p. 129). This is Evans-Pritchard's starting point too, following Durkheim: the proximate and determining cause of society is this fact of aggregation and combination. This is a phenomenon at a different level to that of individual minds; there has then to be a break between psychology and sociology, for the former cannot explain the latter.

The fact of association, Durkheim states, is the origin of all other obligations (1982, p. 130), and these are quite distinct from psychological states. He contrasts, on the one hand, the feelings a man experiences when confronted with forces superior to his own with, on the other, the institution of religion with its beliefs and practices, so multifarious and complicated, and its material and moral organization. Likewise, he contrasts the psychical condition and sympathy between two people of the same blood with the hotchpotch of legal and moral rules which determine the structure of the family, personal relationships, and the relationships of things and persons. Frequently, he adds, social scientists take well-defined and specific psychical states which are the consequences of social phenomena as the causal conditions of the same. In this way, a certain religiosity is used as an explanation for religion, or sexual jealousy for marriage, or filial piety and fatherly affection for the family. Yet under different conditions, these mental states may be absent, or vary enormously. Sociability itself, he suggests, may be learnt from social life (pp. 131–2).

Although he claims to adopt a 'naturalistic' approach, then, Durkheim sets the idea on its head. To offer a summary: the conventional story (to be rejected) suggests that almost-human apes meet to mate, so fulfilling biological needs. Out of these family groups there emerges the first division of labour, along lines of gender and generation, making units bound together by psychological bonds of parental affection and filial duty, arising out of desire and dependency. These are the first humans. Once these family groups begin to interact with other similar groups by extending these familial sentiments of trust and cooperation to their neighbours, the possibility of trade and the further division of labour emerges and with it, society, with a subsequent history of increasing specialization, interdependence and the rise of civilization. Biology underlies psychology, and psychology underlies society. Durkheim reverses this account by enumerating an ordered series of logical terms in the place of a speculative narrative of imagined past events. He suggests that first in logical terms comes social classification, in its simplest form, the division of people into two marriage classes, the people you may marry, and the people you may not: from a male point of view, potential wives and kin. From this social division, sentiments arise, for one feels differently towards marriageable persons than towards kin. In certain cases, these feelings may be given expression in the act of mating. In this account, the social fact precedes the psychological, and the psychological the biological. The naturalistic account has been replaced by the sociological, the individual focus by the collective, the evolutionary time-line by the historical. Psychological states are symptoms of social facts rather than their cause.

As a final proviso, Durkheim points out that 'the psychical element is too general to predetermine the course of social phenomena' (p. 132), and since it does not imply favouring one social form over another, it cannot explain any such forms. We may note that this argument recovers any putative simpler societies, previously left to one side for the sake of the argument, within the remit of sociology. Durkheim concludes on the one hand that the determining cause of a social fact must be sought among the antecedent social facts, and not among the states of the individual consciousness. And on the other hand that the function of a social fact—its purpose or function—must be sought in social form: it produces socially useful effects, not biological or psychical effects (p. 134).

Having established the autonomy of the social fact, Durkheim is more conciliatory in his preface to the second edition of the *Rules* [1901]. First, he restates the case: psychology is concerned with the science of the individual mind, while sociology is concerned with the whole and not with the parts, so sociological facts lie outside the consciousness of individuals and outside psychical facts; although they too consist in ways of thinking and acting, they have a different substratum. The states of the collective consciousness are of a different nature from the states of the individual consciousness: they consist in representations of another kind, representations which change with the state of society, and which create sentiments of loyalty and disapproval (1982, pp. 39–40). Durkheim can in this way distinguish between the psychology of the individual mind and the consciousness of self, which is socially created (p. 36). This argument is presented with great clarity in 'Représentations individuelles et représentations collectives' (Durkheim [1898] 1996).

Once this division is recognized, he can then propose a 'formal psychology' or a 'social psychology' (Durkheim 1982, p. 41), which looks at the way the two kinds of representation work, and whether there are parallels in these workings with regard to proximity and simultaneity, contrast and logical opposition. At present, he states, little is known. Concerning individual representations, we have the association of ideas; concerning collective representations, more or less nothing is known, though Durkheim points to mythical themes, legends and popular traditions, and languages as likely areas for research. 'If psychology were more advanced it would doubtless establish that each category of mental states has its own formal laws that are peculiar to it' (p. 42). Durkheim's contemporary François Simiand went further in exploring such a social psychology, in his investigation of the limits of neo-classical economics. The major point, however, remains that the social psychological mechanisms at stake are aspects of collective representations and their working, and so an expression of sociological concerns.

The Construction of Cognitive Science

With respect to anthropology, Durkheim defined the social field by his identification of the topics of obligation, holism, and collective representations. However, each time there has been an advance in psychological theory, new claims have been advanced of the possibility of explaining social facts in a naturalistic perspective. Durkheim replied to a psychology based on the association of ideas; there have been claims put forward subsequently on the basis of behaviourism, and then of cognitivism, this last drawing first on the notion of the computer as a model for the processes of the human mind, then on the application to these processes of neo-Darwinian thought. Each time, anthropologists have mounted a defence of the autonomy of the social, on the basis of Durkheim's account; first, Lévi-Strauss (1949) and Needham (1962) in response to earlier claims of psychological theories; later, James (2003) and Laidlaw (2007) offer replies to cognitivism. The history of anthropology is closely involved with the history of psychology. Rather than attempt to sketch this history in any detail, I shall outline the situation of cognitive science with respect to its psychological predecessors, drawing on the account to be found in Descombes (2001), in order to prepare to appraise its claims with respect to sociology in the second part.

What is cognitive science? We may identify two features to orient ourselves, considering cognitivism, the philosophical position that permits cognitive science (for this puts us on a level that matches the abstraction of Durkheim's approach). In the first place, cognitivism separates itself from behaviourism, a theory of the 1940s and 1950s. Behaviourism seeks to explain the behaviour of people without appeal to their mental life: it is claimed that there is no need of a concept of mind to mediate between the stimulation provided by the environment and the responses of the organism (see Descombes 2001, p. 3). Cognitivism is, on the contrary, a defence of the notion of mind; it is the study of the mental life of the individual subject or cogito. In contrast to behaviourism, cognitivism places mental phenomena such as language and meaning primarily in the head, and only relates them in a secondary fashion to what happens outside the head. Behaviourism treats words, gestures, writing and so on as input from the environment, and action or behaviour as output (which may be in the form of words, gestures and so forth); it neglects any form of interior life. Cognitivism places mind as an additional layer, mediating between stimulus and action, and highlighting the contrast between inside and outside the head. In this regard, it represents a revival of a dualist account of human nature and understanding, balancing the contributions of mind and body.

In the second place, although cognitivism is a dualist philosophy of mind, unlike many earlier versions, it does not see the subject as the origin of our meaningful utterances. Cognitivists are strict materialists; mental life is conceived of as a physical process, and the mind is to be explained as a material system, the brain (see Descombes 2001, p. 4). In this respect, cognitivism resembles associationism, the psychological theory which preceded behaviourism: both cognitivism and associationism see the problem of mind as one of a mental mechanics. The model of explanation is that of a natural science. In short, cognitivism may be characterized by the two claims, that the mind is inside the head, and that the mind may be explained by the methods of the natural sciences.

The two psychological theories touched on in this brief outline, behaviourism and associationism, serve to bring out respectively contrasts and resemblances with cognitive science. They have both had a role in the history of anthropology, too. I will begin with the older theory.

The association of ideas: The association of ideas as a mental philosophy goes back at least as far as Locke. Mill tried to set the theory on a scientific footing, in order to defend psychology against Comte as a positive science of mind, and in so doing, to extend the methods of the physical sciences to the study of man and his mores. In section 6 of his *System of Logic* (1842) Mill introduced the idea of 'moral sciences'—a science of mores—to match the physical sciences. This attempt to create a single approach was prior to any distinction between naturalistic and hermeneutic disciplines (Descombes 2001, pp. 30–1).

The associationist account therefore sought to present a scientific theory of psychical causality (as does the later cognitivist account). Just as with the physical sciences, it looked for efficient causes and sought to identify tight chains of events. To do this, the theory identified a set of psychical elements and proposed a small number of simple mental mechanisms to account for the constant emergence of effect B whenever cause A arises. In the case in hand, the psychical elements are ideas, and the mechanism of psychical causation is the property that ideas have to call other ideas to mind, either through similarity or contiguity (Descombes 2001, pp. 86–7).[1] Elsewhere, Descombes notes that the 'laws of association' in fact derive from rhetoric, since contiguity and similarity are stylistic figures by which a speaker develops his position in a direction that has been anticipated in the minds of his audience by the words he uses (note 6, p. 267, referring to p. 168).

We might notice the way in which these associationist mechanisms were adopted by J.G. Frazer in his account of the primitive mind and magical thinking. In practice, Frazer took a mental philosophy—a scientific theory of the association of ideas—to be the natural property of all humans; this is a confusion of a theory with a natural condition or state. It allowed him to attribute to 'primitives' a series of mental mistakes by which they are supposed to generate

[1] We may note a possible third 'law', of association by contrast.

various actions and to explain various events because of their intuitive theories of magical causalities, as if they were philosophers of mind.

There are two distinct problems for the association of ideas taken as a scientific hypothesis (and therefore taken as efficient causes with explanatory power). The first concerns how one gives a plausible psychological explanation of someone's mental life; the second concerns how one explains the effects ideas have on behaviour. Descombes terms the first the 'Sherlock Holmes problem' and the second the 'Brentano problem' (2001, p. 168). To take the first, associationism never succeeds in explaining how a chain of thoughts may be a logical sequence of thoughts. The 'laws' of the association of ideas are incapable of making a distinction between a chain of reasoning and a non-sequitur (ibid—citing Fodor 1987). This problem is approached from another angle by Lévi-Strauss in the late 1940s, who pointed out that, if you stick to the level of efficient causes—an accumulation or chain of events—it is impossible to explain social order. Associationism cannot provide a psychology adequate to the needs of sociology. Common sense, if it considers the question of organization, has recourse either to the notion of a legislator, some intelligent mind at the origin of a social settlement, or to chance and the effects of historical contingencies. If one is looking for a psychology to underwrite these notions, the association of ideas, an empirical psychology, relates to contingency, while its only rival at this level is a rationalism without foundation, creative intelligence without a context. Neither is sufficient to explain order, relationship, and, above all, the persistence of institutions. Lévi-Strauss turned to Gestalt psychology, which exhibited a concern with wholes rather than parts, to serve as a model that would allow the creation of a more satisfactory anthropology (Descombes 2001, p. 52). Association as a mechanism produces a train of thought, but not a rational train of thought (p. 173).

Even if a mechanics of mental association could claim to produce a semblance of mental life, it cannot account for the passage to action (p. 168). This is the second problem. Associationism proposes two laws—of contiguity in experience and resemblance—in order to associate ideas without regard to their representational content. But a mechanism of resemblance is no help in creating a mechanical psychology, for it depends upon an association of content and therefore of meaning. This move appears to appeal to an intelligent mechanism, at a lower level than that of mind, and therefore invokes a homunculus, a little man within the head. Associationist theories were discredited for two reasons: their inability to explain rational thought, and the fact that the mechanisms proposed are pseudo-mechanisms (p. 173). These points are relevant to a consideration of cognitivism, to see how it is thought to have overcome these limitations

Behaviourism: Cognitivism is, on its own account,[2] formed in the rejection of behaviourist psychology, which sought to explain people's behaviour without

[2] Chomsky suggests there were two cognitive revolutions: that of Descartes et al., putting an end to scholasticism, and that of today, abolishing the hegemony of behaviourism—see Descombes 2001, p. 77 and Chomsky 1959.

appeal to their mental life, that is, without appealing to the notion of a mind lodged somewhere between the stimulations provided by the environment and the organism's responses (Descombes 2001, p. 3). From one angle, behaviourism might be thought to fit well with Durkheim's claim of the independence of sociological phenomena from psychological, for it brackets the problem of the individual's mind. This is, however, largely an Anglo-American debate: the psychology of autonomous mental life versus the psychology of external behaviour (Descombes 2001, p. 8).

For the cognitivists, the phenomena of mind are indirect; we observe behavioural phenomena and seek their satisfactory explanation. The behaviourist answer would be the right one if we could explain things without having to postulate a 'black box' within the system, lying between the external stimuli and the visible reactions. But behaviour is more complex than can be explained by the control exerted on the organism by the external environment, and we therefore have to have recourse to a 'black box' to generate the observed behaviour (Descombes 2001, p. 18). Descombes points out that cognitivism shares the same conception of observable behaviour as behaviourism, while having the same conception of cause as an earlier mentalist philosophy; it attributes causality to a mental process manipulating representative ideas. Its novelty is to insist these ideas are not ideal constructs, but real cerebral entities (p. 19).

Behaviourism fails, according to Descombes, because it has proved impossible to treat behaviour without applying forms of teleological explanation (he points to Taylor 1964). He calls attention to the debate in the philosophy of action over causes and intentions with respect to the structure of action (see pp. 32 ff). Causalist accounts demand that explanations of behaviour in terms of aims (i.e. teleological explanations) are translated into explanations by means of antecedent causes. Otherwise, it would have to be admitted that something which does not yet exist—a desired particular future state of affairs—can produce effects. Teleology is explained as a shorthand way of appealing to the causality of representations: actions seemingly carried out in pursuit of a goal are in fact carried out under the effect of a previously formed representation.

From the other, intentionalist, point of view, the notion that an aim could be the mental cause of an action is incoherent, on two grounds. In the first place, an intention does not have to be formed explicitly prior to an action, for often we carry out movements intentionally without having first planned to do so. In the second place, it is a (category) mistake to conceive of intention as some kind of event or mental episode. An aim—to write a book, for example—is not the mental act of a particular instant, which causes subsequent conduct. It is rather an ordering of efforts and decisions, of means to an end (Descombes 2001, p. 33)—a structure or ordering of behaviour.

Behaviourism is an attempt to eliminate all teleological or purposive language from the explanation of behaviour. But in practice, the behaviourist simply hides the teleology of the object under consideration, by, for example, shifting it into the experimental situation. Indeed, Descombes suggests, the concept of psychological behaviour is inescapably teleological: the behaviourist eliminates terms such as belief or hope (the rat believes it will find sugar at the end of the path, or hopes to escape from the maze), but includes terms such as seeking (sugar) or avoiding (shocks) (pp. 34–5). If the behaviourist eliminated all intentional language, animal psychology would be left with nothing to study (p. 35). Behaviourism failed because it was unable to naturalize the intentional: behaviourist theories were unable to translate psychological concepts into simple types of overt, and therefore external, behaviour (p. 69). And, the other way about, a behaviourist has to deny that people have opinions and desires (p. 74).

Another critique is elaborated by Fodor, who criticizes behaviourists for refusing to consider cognitive mental processes. These are exemplified by the methods of Sherlock Holmes: a detail makes Holmes think of a certain circumstance, which leads to an idea, and so forth; Watson offers a summary of Holmes' train of thought (Descombes 2001, p. 169). This train of thought needs to be a form of reasoning, an argument that is more or less convincing. This criticism applies equally to the mechanical psychology of associationism, which is also unable to provide an account of the rationality of such episodes (pp. 169–70).

As Chomsky suggested, behaviourism had a period of success (Descombes 2001, p. 77): the objections of behaviourists had led psychologists to doubt the reality of cognitive processes. They were reassured by the appearance of the computer as a model (p. 67). Some of the criticisms of behaviourism, however, also apply to cognitivism (essentially, the attempts to replace reasons with causes, pinpointed by Laidlaw 2007). It is a challenge to find an alternative to cognitivism without reverting to a version of behaviourism, in which behaviour is directly controlled by the external environment (see Descombes 2001, p. 213).

In brief, cognitivism may be understood as repairing deficiencies found in earlier psychological theories, theories which have both influenced and bear resemblances to anthropological approaches. Cognitivism is a theory about thinking, with an emphasis on rational, propositional thought rather than emotions or imagery. It holds that thinking consists in the manipulation of inner representations, which are complex, structured symbols, and that symbols and manipulations can be described in terms both of process and meaning. Haugeland (2004) points out this account is a version of the classical theory which represents thought as inner discourse. This version has been given legs by a series of advances in artificial intelligence which allows the claim that a machine (a computer) can execute at least a good deal of what thought amounts to, suggesting in this fashion a way of understanding how our own intelligence might be implicated in matter. In Haugeland's words, 'reinvigorated cognitivism offered the first really plausible hope of seeing how

Cartesian dualism [the mind/body problem] might actually be overcome' (2004, p. 255). Can this new theory also meet the objections put forward by Durkheim and offer a way forward for both disciplines? To recall these objections: Durkheim's challenge to psychology is to account for the shared ordering of human desires, ideas and habits, their independence from individual needs and instincts, their compelling nature, and the range of cultural variety over and above the range of human emotions.

PART 2 A CASE STUDY

A Comparison of Two Accounts of Witchcraft

This review suggests that social anthropology and cognitive psychology are mutually implicated in some respects, but in others have developed in response to different questions. Rather than prolong this sibling rivalry, I want instead to look at the implications of each approach, for in practice they imply different objects, methods, and forms of explanation. My aim is to seek a common frame, and to search for the appropriate categories to do so, although my conclusion primarily points to the need for further research. In order to lay out the issues clearly, I will take a specific example, and offer two accounts of witchcraft, the one anthropological, the other from the cognitive science approach.

An anthropological account: The first, anthropological, account is that of Evans-Pritchard ([1937] 1958), who offers not only a description and analysis of Azande witchcraft, but also a particularly careful account of his method or approach. He identifies a trio of topics, witchcraft, oracles, and magic, and says he wishes to study how these practices and beliefs relate to one another, and to show how they form a system. 'Azande do certain things and believe certain notions,' he writes, 'and I have tried to explain a fact by citing other facts from the same culture and by noting interdependencies between facts. Explanations, therefore, will be found embodied in my descriptive account and are not set forth independently of it' (p. 5).

Evans-Pritchard also outlines a particular approach to language, being interested in mapping Azande thought and ideas, and employing English words to stand for Zande notions. We can gloss Evans-Pritchard as follows: there is no such thing as 'witchcraft', some kind of universal phenomenon of which we can find instances all over the world and at different times in history. There is no 'essence' of which we find expressions. Rather, we are interested in a series of indigenous phenomena, and we approach these phenomena which have caught our attention through the native classification or categories. In order to orient ourselves, we use a series of words drawn from our own culture, words with which we are familiar, having an intuitive sense of their meaning (which derives from a particular history), and we apply these in a coarse-grained way to these native phenomena and understandings with which we are concerned. We want to understand certain Zande ideas and behaviour, and by using these English terms—'witchcraft, oracles, and magic'—we begin the process of identification and interpretation. The words are tools, no more, with which we begin our mapping of Zande meanings and actions in this area of activity. Evans-Pritchard adds, 'if anyone cares to designate these (Zande) notions and actions by terms other than those I have used I should raise no objection' (p. 9). By this approach, Evans-Pritchard broke with intellectualist (Frazerian) explanations, which had asked how a reasonable person could come to think magically, and so brought the inquirer and the native into a common frame, assuming identical mental processes going on inside the head of both parties.

When he offers a description, Evans-Pritchard does two things: he links popular beliefs on the one hand and changing institutions—or micro-politics—on the other. He details the Zande belief system with respect to moral causalities and classes of persons, the attribution of intention, forms of divination and accusation, and processes for reconciliation, restitution, and punishment. In effect, he sketches out an account of an idiom of 'neighbourliness'—of good neighbours and bad—which can be mobilized in particular under conditions of social change and disruption.

This is the second aspect of Evans-Pritchard's analysis, the effects of institutional change on the deployment of this idiom of neighbourliness. He points to changes effected under colonial rule—bringing unrelated people into closer proximity through resettlement and, at the same time, abolishing the powers of native courts, so that at the same time as accusations increased in frequency, the ways of dealing with them were overturned. These factors help explain why at the time of his visit, witchcraft events were prominent in a way they may not have been at other periods.

This account offers a clue as to how we may detect parallel phenomena and make comparisons without recourse to essentialism, that is, without assuming a common idea (witchcraft) is being manifested in two distinct contexts. The most important feature of the description concerns the interaction of moralities and institutions: how social change is received and interpreted in terms of folk categories, and how it may, particularly through changes in law, generate unintended consequences. That is, in brief, what European witchcraft is: when 'bad neighbours'—a folk category—can be identified as occupying a new legal category of unorthodox belief, associated with political sedition. We are not dealing with an essence but with human constructs which, at best, share certain 'family resemblances', resemblances which can be extended to point to comparable phenomena in other settings.

A cognitive science approach: Let us now consider a cognitive science approach to the same topic, by reading Cohen's recent article 'Witchcraft and sorcery' (2007). Cohen is concerned with 'the information processing devices of the human mind', devices that are held to have 'evolved in response to selective pressures in our ancestral environment' (p. 136). This inheritance may under certain conditions generate human cultural practices such as 'religious ideas, social relations (and) moral prescriptions'. These devices allow the possibility of such beliefs, and so allow some sort of explanation of witchcraft practices. This universal heritage allows us to explain on the one hand resemblances between widely distributed human practices (instances of witchcraft belief) and, on the other hand, the persistence of these practices, their transmission between generations and between populations.

The pervasiveness of witchcraft beliefs is taken as a natural fact that demands explanation. And 'parochial accounts', such as that offered by Evans-Pritchard, 'must give way to, or at least be complemented by, accounts that can explain the recurrence of these beliefs across cultural contexts and throughout history' (Cohen 2007, p. 137). We should note, then, that Cohen offers a different account to that of Evans-Pritchard at each level of the project, concerning the use of language, the nature of the project of explanation, and the possibility of comparison. Cohen offers an account in which words name natural objects and the effects of natural objects; this function indicates the object of the project, the specialized cognitive structures that favour the spread of beliefs in witchcraft and spirits; and the constancy of this object allows comparison as a straightforward process.

The most notable difference is these cognitive structures are not designed to produce religious beliefs; they relate to a far wider range of perception and representation of the world and only incidentally produce the beliefs that concern us; witchcraft beliefs and practices are the unintended and hence motiveless by-products of more broadly focused adaptive information-processing devices. So while an anthropological approach seeks to identify a logic or rationale to witchcraft practices, Cohen denies the possibility of there being an inherent rationale of the kind given by Evans-Pritchard.

Cohen considers two mental mechanisms, first in their broad working and then bringing them to focus on the topic of witchcraft practices. In the first place, there are 'attribution theories', whereby people attribute agency as the causes of events in the world (2007, p. 138). Humans, it is claimed, display a 'naïve' biology, physics, and psychology, each relating to a different mental domain. The third of these devices, 'the mechanisms that enable us to make inferences about other people's intentions' (p. 140), is commonly termed 'Theory of Mind' in the literature: we impute the possession of minds to other creatures like ourselves, and therefore intentions. And in the second place, there is a class of 'essentialist' reasoning, whereby members of a class are perceived as possessing some inherent quality which simultaneously defines membership of a category and causes certain typical properties of the category to emerge (such as appearance and behaviour). The first looks to intentions, the second to inherent essences, as active in the world; Cohen looks to the former to explain cases of 'sorcery' and the latter cases of 'witchcraft'. But these are, to repeat, only side-effects: 'both sets of mechanisms', Cohen observes, 'form part of ordinary cognitive equipment that is available to all human beings and that is crucial for inferring, acquiring and organizing information about the social environment' (p. 138). She adds that, because both intention and essence are by their nature hidden, ritual means—such as divination and exorcism—have developed to discern and deal with them.

Let us note briefly a number of significant features of the theory. The first is that a good deal of the relevant evidence comes from developmental studies carried out on young children. Second, while 'religious' beliefs such as gods, spirits, and sacred objects are taken to emerge through the misapplication of social cognitive devices to events generated in the world, it can be argued that these category mistakes have a selective value. Third, while these ideas appear spontaneously, they spread because they are, in Cohen's words, 'constrained by tacit, intuitive ontological knowledge about the world' (p. 145). This account is elaborated further in terms of their spreading because they are memorable, being 'minimally counterintuitive' notions, possessing familiar features with a single exception: persons without physicality, such as spirits and ghosts, for example. Together, these ideas point to the importance of the development, stability, and communication of ideas.

Fourth, Cohen is concerned to explore how the two domains—Theory of Mind and essentialist reasoning—interact. Although we cannot follow this path, it introduces considerable complexity into the account. Fifth, ideas begin in an atomistic fashion, and are shared between individuals because of compatibilities in the structure of minds; extension is an important postulate in the construction of the social, which only enters at this stage. This is in distinction to Evans-Pritchard's concern with the primacy of classifications as collective representations. Last, the approach argues that a series of utilitarian features—an atomistic psychology, a functionalist teleology, and a naturalistic essentialism (or, in English, the individual, endowed with intentions or ends, since first becoming man)—is the product of selective evolution, and therefore may be used to describe the innate, shared properties of the human mind.

The error at the root of popular beliefs concerning witchcraft and exorcism is taken to concern misapplying this capacity to perceive others as members of natural categories and imputing to them a non-observable causal essence that produces a manifold of observable characteristics and events. So when Cohen considers Evans-Pritchard's work, she focuses on the inherent quality of witchcraft substance, the indeterminacy of the state of being a witch, and the need for divination. She develops the category of the witch by extension from the individual to close kin, forming in this way a quasi-natural category. And she builds on this definition by bringing in material from other cases of 'witchcraft', first from Africa, then from Indonesia, then from seventeenth century Europe, to develop the description. While not everybody entertains the possibility of a 'witch' category, she concludes, where they do, the category is the product of normal cognitive resources for parsing the social domain.

This approach is scientific, Cohen states, dealing with natural dispositions, triggered by features of the environment, and offering testable explanations. Its apparent objects are agents, beliefs and essences, but in practice, these are just 'noise' generated in the margins of the real object of the cognitive science of religion, the workings of the mechanisms of the evolved human mind.

The two approaches compared: The two approaches may be distinguished in a number of respects. They differ as to their object, methods and principles of explanation. Notably, they employ quite different scales both in terms of extent and time: the units they study are quite distinct. And underlying these differences there are different logical forms, the one allowing overlap of kinds, the other relying on identity and non-contradiction.

To characterize Evans-Pritchard's approach, mind is outside the head: it is held in the interactions, interpretations, common assumptions, and collective categories of a diverse group of people. His approach is therefore holistic, and his objects include, on the one hand, systems of classification and, on the other, events. His methods then consist for the most part in mapping simultaneously systems of classification and behaviour, recognizing the interdependence of facts and interpretation, and so treating language as a part of reality, for the registration of an event is an integral part of its being an event.

Explanation is then embodied in description, and this form of explanation is inescapably historical, working with a nesting set of contexts and events. It is worth emphasizing the narrative construction of understanding, not least because, in such a perspective, any psychological account presumes such a narrative. Mental states arise out of particular pasts, and are held against some future horizon; without these, their present does not exist.

The possibility of comparison between non-identical cases and comprehension arises from the reconstruction of overlapping forms, epitomized by Evans-Pritchard's attitude to language: there are certain terms which we understand from our own history which give us a first orientation with respect to the indigenous events to which our attention has been drawn. A great deal of work is then needed in order to complete that work of reconstruction. There is no intuition of native forms, nor an assumption of any common identity of meaning. In short, Evans-Pritchard's programme is this: mind outside the head, a mapping of indigenous categories which includes events, and a situated, historical account that escapes from a logic of identity and non-contradiction.

And time impinges in a series of relatively short scales: there is a longer-term horizon of Azande 'traditional' practices combined with the shorter-term impact of colonial practices, which in turn generate short-lived events of witchcraft accusations and confessions. These are monitored by Evans-Pritchard, who begins to make sense of what he finds with some of his own cultural terms passed through various filters: these arise in sixteenth and seventeenth century European materials and are transmitted through subsequent recensions.

Let us now characterize the cognitive science approach offered by Cohen. It is worth remarking that, although cognitive approaches reject anthropological explanations, they rely for the main part upon the anthropologist's initial description of the phenomenon, and this is because the object of the approach is a secondary product, a re-description or reinterpretation of materials. For Cohen, the mind is inside the head. The initial object of study is therefore smaller in scale than the anthropologist's, for it concerns the individual and the range of processing devices in the individual's brain. Yet because these devices are the products of evolutionary processes, the scale is also far vaster, being, indeed, universal in the sense of being a part of the biological apparatus of all humans. So the scale is both smaller and larger than in the sociological case, and the objects are not patterns of observable human behaviour, but inferred, unobservable constructs. This is why re-description is the issue: the specific causes of witchcraft accusations are secondary, as are reasons for the absence of accusations under other conditions. What is of primary interest is the mechanism that underlies the possibility of such expressions.

This focus upon mechanism is supported by a particular account of the human mind and its out-workings. Mechanisms such as the Theory of Mind and essentialist reasoning are expressed in the form of beliefs. Beliefs are individual in origin, and precede action; they are expressed in feelings and behaviour which in turn creates relationships. They do not however remain individual, but get passed from mind to mind; criteria of ease of transmission become important, therefore, since some beliefs are favoured by shared mental mechanisms. In this way, too, social relationships develop by extension of beliefs and sentiments to classes of persons, the classes themselves being favoured by other mental devices. We should remark that sociological critiques tend to focus on the extrapolation of sentiments from individual to class relationships, because such relationships appear to change their nature and function in such a transformation, and the mechanism of extension is unclear.

The primary object is then the natural individual considered in the perspective of an atomistic psychology. We should note the lack of any overt reference to the observer. Various methodological features follow, notably a shift in focus from the social setting to detecting the beliefs and motivations of the individual subject, remarked as a concern with re-description and interpretation, and recourse to studies of child development, for these give access to the various mechanisms as they appear in the processes of growth and maturation.

Because of the focus on intentions and essences, both of which are hidden, explanation at one level means making the hidden visible. In this account, ritual becomes some kind of displacement activity, whereby men revisit their category mistakes and repair their consequences. 'Events' do not have the primary importance the anthropologist attributes to them, but instead become individual acts of attribution of significance to movement in the world. Likewise, the business of classification becomes an individual process, to be shared subsequently. Explanation takes place rather at the level of the identification of mental devices through their effects, and the interpretation of interactions between these devices. This account then does not seek to resolve the two issues identified earlier in the consideration of associationist thinking; it is concerned neither with the rationality of thought processes, nor with how thought is translated into meaningful action.

Other features share in this substitution of a naturalistic for a historical perspective. Notably, words are taken to name natural objects, as if the objects were distinct natural species that could be classed in separate genera. In this account, comparison between cases of 'witchcraft' is unproblematic, for witchcraft is a natural object and other ethnographies can be readily employed to supplement gaps in an existing account. The possibility of describing family resemblances and exploring overlap of forms disappears in this approach through the recourse to criteria of identity and non-contradiction. And native uses of language which employ resemblance and overlap stand in need of correction, just as do ethnographic descriptions which map them. Indeed, the category mistakes that are taken to underlie beliefs in gods and ghosts may be paralleled by the anthropologist's creation of systems of meaning and a theory of events.

Rather than dwell on the question of error, however, the feature to which I wish to draw attention is that of scale. Just as the human scale of the two approaches is quite different, the one concerned with both the singular and the universal, the other with a series of middle-distance views, so is the time-scale. Time is not absent from the Cognitive Science approach, but it is simultaneously very long-term and very short. Evolutionary time is taken to be over hundreds of generations and, as such, conjures away any history of events. On the other hand, perception of evolutionary time depends upon a specific short-term event, the advent of neo-Darwinian cognitive science which, only now, can perceive the true object of a scientific psychology, the individual human mind and its development.

The Possibility of Reconciliation—A Darwinian Coda

These two approaches scarcely impinge upon one another. Can they be brought into a common frame? The crucial issue is the relation of history to nature, and therefore the scale of evolutionary time. My grounds for paying attention to this relation come from Collingwood's essay on *The Idea of Nature* (1945). Collingwood suggests that 'the self-consciousness of man... provide(s) a clue to his thoughts about nature' (p. 13), and he traces this changing relationship through three forms. First, he cites the Greeks, for whom 'it had been an axiom that nothing is knowable unless it is unchanging'. Then he considers Renaissance thought, which sought the unchanging true objects of natural science in matter and natural law: in the substance to which the changing arrangements and dispositions of secondary qualities applied, and in the laws according to which these changes could be grasped (p. 11).

Scientific accounts still struggle with notions of matter and laws but, for Collingwood, the crucial discovery of the last two hundred years is a scientific history which has learnt to think about 'a world of constantly changing human affairs in which there was no unchanging substrate below the changes, and no unchanging laws according to which the changes took place' (pp. 12–13). This discovery, 'the historical conception of scientifically knowable change or process was applied... to the natural world' (p. 13), under the broad label of evolution. In short, the concept of evolution is possible because of advances in historical thinking.

As I mention in my introductory remarks, a number of anthropologists have mounted a defence of the autonomy of social facts in the face of naturalistic claims by new forms of psychology. Amongst these, I want to draw attention to Wendy James' *The Ceremonial Animal* (2003), for while she shares the Durkheimian focus on the collective human construction of meaning, she also engages positively with contemporary studies in evolutionary biology. She considers advances in genetics, medical applications, and the mapping of the human genome, pointing out that exploring gene expression in terms of protein production both modifies 'the more rigid views of genetic determinism' and reintroduces the notion of the environment, considered both in terms of nurture and of the social context of learning. She also pays attention to research in ape behaviour and in the development of human characteristics, studies which emphasize the partial and fragmentary nature of evolution, with characteristics emerging in different places and at different times, and likewise pointing to the complexity of group living as a crucial driver in the emergence of such features as brain size and language use. In brief, contemporary 'naturalistic' accounts of human origins are thoroughly seasoned with collective sociological factors from the outset. James suggests, on the basis of this shared symbolic consciousness, that 'to reduce the symbolic complexity of present human action to models of individual rationality, or to examine items of culture and language outside their connection with social form, are bound to fall short' (2003, p. 33).

Since we have just passed the Darwin bicentenary year, let me then conclude by proposing the possibility of reconciling a naturalistic and a historical account by historicizing our notion of evolutionary processes. For the principal ground which divides Evans-Pritchard's account of witchcraft from Cohen's is that of duration. The historical or anthropological account integrates a series of intermediate time scales in a scale of forms which involves both an alternative logic (of overlapping identities) and a method of investigation and explanation through reconstruction. The naturalistic account conjures away these time scales by assuming an extremely long evolutionary process summed up in a self-conscious present, the truth of evolution grasped by cognitive science.

My suggestion is that contemporary evolutionary theory no longer assumes a more or less uniform genetic profile to a vast population united by common descent; instead it emphasizes small populations, interacting in different environments with a variety of partners, so that (to offer an imaginary instance), in one valley, orchids and wasps are involved in symbiosis, in another, interaction is with other species of insect—flies, beetles, or ants—and in a third, competition with another plant species shapes the local outcomes. Just as the study of genetic expression has moved from gene to protein, and that of human capacities has moved from brain to mind, so the study of evolutionary processes has moved from large uniform populations defined by common descent to small heterogeneous populations involved in diverse local alliances across species (or 'against nature'). In a word, naturalism is becoming historicized.

A crude naturalism tries to return to earlier notions either of matter with an inherent tendency to progress, or of natural laws, whose location is unclear, but which likewise are cast in a teleological idiom. On the grounds Collingwood outlines, and on the basis of the evidence sketched above, it is possible to anticipate the development of a social psychology which may reconcile some of the present oppositions. But the reconciliation will be through making nature historical, rather than making history natural, at the cost of the human.

REFERENCES

Chomsky, N. 1959. A review of B.F. Skinner's 'Verbal behavior'. *Language 35*(1): 26–58.

Cohen, E. 2007. Witchcraft and sorcery. In *Religion, Anthropology and Cognitive Science*. Eds. Whitehouse and Laidlaw. Durham, N.C.: Carolina Academic Press, 135–60.

Collingwood, R. G. 1945. *The Idea of Nature*. Oxford: Clarendon Press.

Descombes, V. [1995] 2001.*The Mind's Provisions. A Critique of Cognitivism*. New Jersey: Princeton University Press.

Durkheim, E. [1895] 1982. *The Rules of Sociological Method*. Basingstoke: Macmillan.

——. 1982. 'Preface' to the second edition of the *Rules*. In Durkheim, 34–47.

——. [1924] 1996. Représentations individuelles et représentations collectives [*1898*]. In *Sociologie et Philosophie*. Reprint, Paris: Quadrige/PUF, 1–48.

Evans-Pritchard, E. E. [1937] 1958. *Witchcraft, Oracles and Magic among the Azande*. Reprint, Oxford: Clarendon Press.

Fodor, J. 1987. *Psychosemantics: The Problem of Meaning in the Philosophy of Mind*. Cambridge, MA: MIT Press.

Haugeland, J. 2004. Closing the last loophole: joining forces with Vincent Descombes. *Inquiry 47*: 254–66.

James, W. 2003. *The Ceremonial Animal*. Oxford: Oxford University Press.

Laidlaw, J. 2007. A well-disposed social anthropologist's problems with the 'Cognitive Science of Religion'. In *Religion, Anthropology and Cognitive Science*. Eds. H. Whitehouse and J. Laidlaw. Durham, N.C: Carolina Academic Press, 211–46.

Lévi-Strauss, C. [1949] 1969. *The Elementary Structures of Kinship*. London: Eyre and Spottiswoode.

Mill, J. S. [1842] 1970. *A System of Logic Rationative and Inductive*. London: Longman.

Needham, R. 1962. *Structure and Sentiment: A Test Case in Social Anthropology*. Chicago: University of Chicago Press.

Taylor, C. 1964. *The Explanation of Behaviour*. London: Routledge & Kegan Paul.

Whitehouse, H. and J. Laidlaw. 2007. *Religion, Anthropology and Cognitive Science*. Durham, N.C: Carolina Academic Press.

11

Religion: The Dynamics of Cultural Adaptations

LESLEY NEWSON AND PETER RICHERSON

The goal of science is to achieve a level of understanding which not only explains current observations but also allows us to make predictions. Can science provide this kind of understanding of religion? Can it allow us to predict how holding beliefs categorized as 'religious' is likely to affect people's lives, how these effects are likely to play out in the future or how religion itself will change? Despite its interdisciplinary nature, research in cognitive science cannot by itself provide this kind of practical understanding. In this chapter, we will argue that a broader approach, which includes insights gained through studies of cognition, can allow us to develop theories of how changes in the social environment people experience bring about religious change and changes in the values that govern their lives.

Cognitive scientists observe and develop theories about how individual brains process information. However, as Purzycki, Haque, and Sosis (this volume) suggest, gaining a practical understanding of how human minds process information requires that we take into account that fact that humans process information *socially* as well as within their own individual minds. The human brain is equipped with an array of psychological mechanisms which allow us to share the processing of information during social interactions. As children we acquire a large repertoire of cognitive skills, emotions, attitudes, beliefs, and behaviours from the older people who care for us, not just as a result of the genetic endowment from our parents. As we go through life we observe our environment in the presence of other minds. We observe people around us and are aware that they can observe us. We use our unique communication system, language, to discuss perceptions, behaviours, the mental representations that appear to motivate behaviours and many other aspects of our shared lives. The individual and social information-processing mechanisms are intricately interconnected and continuously feed back on each other. We share thoughts and create ideas *together*. We are often not aware if our beliefs about an event are based on our witnessing of the event or the stories told by our fellow witnesses (Loftus 1996).

In this chapter we argue that, to gain an understanding of how minds develop the beliefs, values, and behaviours categorized as 'religious', it is necessary to take into account that this development occurs in minds that are situated within social networks. Our argument begins with an explanation of why we see cultural change as a 'Darwinian process'. We then proceed to a review of recent work in comparative primatology and human development, which suggests that children are adapted to acquire a vast amount of relatively unconstrained information by imitation and teaching and that humans are able and motivated to 'change their minds' throughout their lives. Like all aspects of human culture, religious beliefs can be thought of as being 'held' both by the community and the minds of the community members in which the information is actually experienced and stored. Like all aspects of culture, religious beliefs and customs don't remain static. The culture of a population changes as its membership changes, through birth, death, migration, and experience. Members are constantly learning and sharing new information. New beliefs, customs, and rules are 'born' while old ones are abandoned, forgotten, or modified (Boyd and Richerson 1985).

Centuries of scholarship have been devoted to developing explanations of cultural change, identifying the events, individuals, and environmental factors that might have influenced its course. But in recent years, a growing number of scholars have begun to develop broad theories of cultural change with the help of mathematical modelling tools similar to those used by population geneticists to understand the genetic changes that occur in a population (e.g. Boyd and Richerson 1985; Cavalli-Sforza and Feldman 1981; Joseph Henrich 2004; McElreath, Boyd, and Richerson 2003; Newson, Richerson, and Boyd 2007; Richerson and Boyd 2005, 2010). This amounts to the revival of an approach outlined by Darwin (1874) in *The Descent of Man*, and which was pursued by a number of prominent pioneers in the social sciences, including William James. However, for historically contingent reasons Darwinian ideas virtually disappeared from the social sciences around the beginning of the twentieth century (Hodgson 2004; Richards 1987). The applicability of these ideas to cultural change was rediscovered in the mid century by a number of authors (Alchian 1950; Campbell 1960; Gerard, Kluckhohn, and Rapoport 1956). Sophisticated theoretical models of cultural evolution were pioneered by Cavalli-Sforza and Feldman (1973) and we are currently in the midst of an explosion of empirical work (e.g. Henrich and Henrich 2010; Mesoudi 2007; Whiten, Hinde, Laland, and Stringer 2011).

Darwinian theory is often thought of as one of those grand theories of everything of which most modern social scientists and historians are properly sceptical (Slingerland 2008). It should instead be thought of as a large tool-kit of models and empirical approaches which can be deployed flexibly to solve a wide range of problems (Richerson and Boyd 1987). This tool-kit is useful whenever a problem involves transmitted information changing over time. It is easy, for example, to use Darwinian theory to show how historical contingency creates cultural (and genetic) diversity (Boyd and Richerson 1992). We have used this approach to look at the pattern of cultural change sometimes referred to as 'modernization', which is associated with economic development (Newson and Richerson 2009). These changes include changes in religious belief.

In this chapter we give a non-mathematical summary of the cultural evolutionary approach to explaining cultural change and the evolution of human behaviour, and show that this provides us with a novel view of the role of changing religious belief in modernizing societies. We review the evidence that the size and connectedness of a population determine the complexity of the culture that a population can support. We then look at the changes in the size and connectedness of populations which have occurred in the last 200 years as societies have undergone economic development. Change in the structure of societies is accompanied by a process of cultural change which is broadly similar in most but not quite all populations (Newson and Richerson 2009). Finally, we look more closely at the pattern of change in religious belief that accompanies this process.

'COGNITION-LITE' AND THE EVOLUTION OF CULTURE

For many, the word 'evolution' implies progression—progress from the less complex to more complex. Darwin (1859) didn't see evolution in this way. In fact, he didn't use the word 'evolution'. He described a process ('descent with modification') by which the variety of life on Earth, both complex and simple, could be the result of a long and continuing process of change. Darwin noted that even though groups of organisms may be similar enough to be categorized as belonging to the same species, individual variation can still be detected. Offspring are more similar to their parents than they are to the population at large; descendants inherit some of the characteristics of their parents. Characteristics associated with individuals who survive and produce more offspring will therefore be more common among the offspring generation than the parents' generation. In the course of a number of generations this 'natural selection' of characters can transform a population. In this way, a population of organisms considered to be a single species can diversify into two or more species if it is divided into groups that occupy environments which favour different characteristics.

Since Darwin's time, biologists have gained greater and greater understanding of the mechanisms by which variation is inherited. If variation in the population is due to differences in genetic information (DNA), individuals resemble their parents because their genes are copies of their parents' genes. During the twentieth century most scholars who used an evolutionary approach to understanding human behaviour concerned themselves with speculating about how variation in this genetically stored information might result in behavioural variation. But the information stored in our genes is not the main source of variation in human behaviour. Behavioural variation is far more likely to be the result of differences in the information stored in our *brains*. Each individual has had a unique set of experiences and they have processed the information gathered during these experiences in a unique way.

The information in our brains is clearly not 'inherited' in the sense that it is copied from the information possessed by our parents. But it is equally clear that much of the information in our brains is very similar to information held in the brains of our parents and the other people we have associated with. We usually speak the same language as our parents, for example. We may have our own way of phrasing and pronouncing our speech but our vocabulary is (and has to be) virtually identical to the people we interact with. The mental representations that underpin our behaviour have to be similar to those of the people we interact with or we could not coordinate our behaviour. This is why the information stored in our memories cannot simply be the result of our own independent processing of our own independently acquired information.

We not only learn a great deal from other people, we also rely on other people to help us process information from our environment. Humans' highly developed capacity to make use of learning from one another and learning *with* one another is one of the things that set us apart from other animals (Tomasello 1999). The ability and the motivation to share learned information results in human populations developing a complex culture and this capacity has made our species highly successful (Boyd, Richerson, and Henrich 2011; Whiten, Hinde, Laland, and Stringer 2011). A young human born today can expect to benefit from a legacy of information that has been built up over thousands of years, and continues to increase in diversity and complexity at an increasingly rapid rate (Carey 2009). When humans are young, a phase of our lives when we acquire many of our most fundamental beliefs, we readily adopt socially transmitted information that is not supported by evidence from our own senses. For example, interviews with children aged four to eight revealed them to have a firm belief in invisible entities, such as germs (Harris and Koenig 2006; Harris, Pasquini, Duke, Asscher, and Pons 2006). On Harris' and co-workers' account, children believe all sorts of things on the basis of the testimony of adults without any direct evidence of their own. They note that

American children are more sceptical about the existence of God than the existence of oxygen, perhaps because they know that adults in their culture have various attitudes about the existence of God but no adults dispute the existence of oxygen, even though few people besides chemists have actually seen persuasive evidence of the existence of the element. On this account the human mind need not have any special cognitive propensity to believe in gods, just a general tendency to trust the evidence of adult testimony. Carey (2009) argues that an early developing 'core cognition' sets up a culturally scaffolded developmental process that allows children to acquire concepts that are quite outside the concepts of core cognition themselves.

Let us call this the 'cognition-lite' picture of how human minds acquire culture. Cognition is fundamental to this picture, but it supports a relatively weakly constrained system of cultural transmission in which entirely novel culturally constructed concepts, such as the scientific discoveries of the last few centuries, are relatively easily transmitted to children. This key cognitive adaptation created a system for producing highly creative solutions to living in almost all of the world's terrestrial environments, including a diverse array of amphibious adaptations that exploit fresh waters and oceans. The cognition-lite picture may be contrasted with the cognition-heavy explanations which rely in detail on innate cognition. This is applied to religion most forcefully by Pascal Boyer (2001):

> In particular, it is clear that our minds are not really prepared to acquire just about any kind of notion that is "in the culture." We do not just "learn what is in the environment," as people sometimes say. That is not the case, because no mind in the world–this is true all the way from the cockroach to the giraffe to you or me–could ever learn anything without having very sophisticated mental equipment that is prepared to identify relevant information in the environment and to treat that information in a special way. Our minds are prepared because natural selection gave us particular mental predispositions. Being prepared for some concepts, human minds are also prepared for certain variations of these concepts. As I will show, this means, among other things, that all human beings can easily acquire a certain range of religious notions and communicate them to others (Kindle locations 52–7).
>
> The idea of "replication" is very misleading. People's ideas are sometimes roughly similar to those of other people around them, not because ideas can be downloaded from mind to mind but because they are reconstructed in a similar way. Some ideas are good enough that you will entertain them even though your elders did not give you much material to work with, and so good again that your cultural offspring will probably home in on them even though you too are an incompetent transmitter! (Kindle locations 542–4).

This is of, course, a quantitative issue. As the mathematical models of cultural evolution make clear, the biases on cultural variants can in principle be so strong as to essentially obviate the influence of culture or so weak as to make culture essentially gene-like in its transmission fidelity and resistance to cognitive remodelling (Boyd and Richerson 1985). Empirically, we can easily make a case for the cognition-lite picture of culture (Boyd, Richerson, and Henrich 2011). A completely novel concept like oxygen and facts about its basic causal properties can be successfully reproduced in any mind which has received a modicum of science education—and in no mind before oxygen's discovery in 1744. Only a tiny scientific elite understands the actual evidence for this particular theory of combustion and aerobic metabolism but 'oxygen' and myriad similar concepts are part of our system of beliefs. This is testimony to the relative freedom of human cultural variants from strong cognitive editing. Of course, religious ideas have a much deeper evolutionary history than oxygen and it remains plausible that relatively weak cognitive biases specific to religion exist, either as by-products of other cognitive functions or as specific adaptation to aspects of religion. Gervais et al. (2011) review this issue in an even-handed attempt to roughly locate religion on the cognition lite-heavy continuum.

In the cognition-lite view, much of the cognitive work in shaping cultural evolution derives from necessarily weak general purpose decision-making systems, with the cognitive workload spread over many individuals and substantial periods of time. Indeed, the cognitive forces often end up acting a bit like natural selection. They influence the adoption of novel cultural adaptations but allow it to still behave like a system of descent with modification (Perreault 2012). Incidentally, Chomsky, whose ideas were highly influential in the development of the 'cognition-heavy' position, has now adopted both an evolutionary and a cognition-lite position (Hauser, Chomsky, and Fitch 2002). Recent advances in evolutionary linguistics have also now taken the cognition-lite perspective (Bouckaert et al. 2012; Hurford 2011).

A cultural evolutionary cognition-lite view allows the development of theories of cultural change which generate testable hypotheses about cultural change. Beginning in the early part of the twentieth century, biologists studying genetic change in organisms making up ecosystems developed mathematical tools to model the change processes and predict long-term effects on population gene pools. The similarity between genetic change and cultural change has motivated mathematically inclined students of cultural change to reshape some of these tools to study the evolution of the culture of human populations (Boyd and Richerson 1985; Cavalli-Sforza and Feldman 1981; Mesoudi, Whiten, and Laland 2004). These are the sorts of tools we have used to try to help explain the phenomenon of modernization and related religious change. Some features of cultural evolution are rather similar to those of genetic evolution. For example, natural selection ought to act on cultural variation much as it does on genetic variation—useful ideas persist and not-so-useful ones die out. But mechanisms of cultural inheritance are different from the passing on of genes from parents to offspring. For example, we were endowed with our parents' genes at the moment of conception but we have inherited our religious beliefs from a host of people and these beliefs are likely to change throughout our lives. Modelling cultural evolution requires modification of the population genetics models to account for these differences but the basic workings of the cultural inheritance system make cultural change ('cultural evolution') a Darwinian process of 'descent with modification'. Cultural change can be seen as being pushed in one direction or another by various forces, which include biases introduced by human cognition. Even if cognitive effects are weak at the individual level they act as strong forces at the population level (Boyd, Richerson, and Henrich 2011).

DEMOGRAPHY AND THE EVOLUTION OF CULTURAL COMPLEXITY

One basic pattern in cultural change which has been observed is the existence of a relationship between the size of a population and the complexity of its culture (Henrich 2004; Kline and Boyd 2010). The larger the population, the more often one of its members will add to their cultural information by discovering a useful piece of information, inventing a new tool or developing some other useful innovation. Also, a larger population is more likely to continue to preserve the useful knowledge and expertise of earlier generations. For a cultural variant to be retained, at least one individual from every couple of generations has to be willing and able to acquire that expertise and pass it on. When a skill or body of information is complex and difficult to learn, individuals with the motivation and cognitive ability to learn them are likely to be rare. In a small population an individual with sufficient expertise may not occur in any given generation. The hard-to-learn skills that are most easily forgotten by a population are also the ones that are hardest to reinvent.

A link between the size of a population and cultural complexity does not just make theoretical sense; it is supported empirically. The best-known evidence is provided by the 4000 or so people living on the island of Tasmania when the first Europeans landed there in 1642. For many years, anthropologists judged the Tasmanian natives to be a quite different and much more primitive people than aborigines living a mere 200 kilometres away in Australia. Unlike the Australians, the Tasmanians had no needles and so they had to survive in their chilly habitat without tailored clothing. They had no boats and no means of catching the fish that surrounded their island. Their technology, which consisted of only 24 tools, was the simplest of any known human group. But when archaeologists began to dig on Tasmania in the 1970s, they found that the people who had lived on Tasmania thousands of years earlier did have boats, fish hooks, needles—the same technology as the people of Australia. It is now known that until 10,000 years ago the Tasmanians and Australians were one people. Tasmania was not an island but a peninsula attached to Australia by a bridge of land. The rise in sea level that occurred about 10,000 years ago drowned this land bridge and the Tasmanians became a small isolated island population. The evidence suggested that at this point the Tasmanians began to gradually lose technology.

More recently, comparisons of the cultural complexity found in the islands and archipelagos in Polynesia have revealed the same effect (Diamond 1978; Kline and Boyd 2010). The smaller and more isolated the population, the simpler its culture. Thus, when we observe the religious beliefs and rituals of a fishing community on a small Pacific island to be simpler than those we observe at the Vatican, we do not conclude that the genes of Europeans endow them with a more complex religious mind. In fact, we know the genes of Pacific islanders to be virtually identical to those of Europeans. Instead we conclude that Roman Catholicism is more complex because many more people have been involved in using and contributing to this body of cultural information.

For many years, however, speculation about the course of human evolution has been dominated by the idea that the size and complexity of human culture can be explained by the size and complexity of the human brain. This is in spite of the observation that hominids who lived 200,000 years ago had skulls that appear to have housed a brain about the same size as modern human brains but have left no evidence of having created complex tools, religion, or art (Shea 2008; White et al. 2003). A small number of more complex human artifacts have been found in Africa dating from between 164 and 40 thousand years ago. But these findings suggest that cultural complexity began to emerge several times, was limited to a single region, and was maintained for only a short time period (Jacobs et al. 2008). All contemporary humans are undoubtedly descended from African emigrants but it was not until about 45,000 years ago in Eurasia that complex culture arose and was sustained. The artifacts left by these 'Cro-Magnon' people clearly demonstrate that they were inventive and capable of symbolic thought. This sudden 'great leap forward' in what archaeologists call the 'Upper Paleolithic' has led to speculation that a genetic mutation occurred which caused the large hominid brain to be reorganized in such a way that its cognitive capacities were greatly enhanced (Diamond 1997; Klein 2009). No evidence of such a genetic change has yet been found, however.

A link between cultural complexity and the size of the communicating population provides another plausible explanation for the time lag between the evolution of large brains and the emergence of complex culture (Powell, Shennan, and Thomas 2009). The paucity of hominid fossils dating from the last two million years compared to that of other animal genera suggests that for most of human evolution, our ancestors were not very abundant (Bobe and Leakey 2009). The larger-brained hominids of the *Homo* genus were widely dispersed and have been found in a range of habitats. This suggests they were able to adapt to making a living in different environments. But the small number of fossils found suggests that they were seldom very abundant until after 40,000 years ago. The Pleistocene was generally a time of great climatic instability but human abundance in Africa did not increase appreciably during the last interglacial period between about 80,000 and 125,000 years ago when the climate was relatively warm and stable for a considerable portion of the time. When humans left Africa about 60–70,000 years ago, and became more abundant everywhere, the climate was at the beginning of a time of extreme variability (Atkinson, Gray, and Drummond 2008; Martrat et al. 2007). The isolated periods of increased cultural complexity observed in Africa may have coincided with times when conditions in an area allowed the human population to temporarily rise. When conditions deteriorated, members of the population were forced to disperse and, like the isolated Tasmanians, they began to lose knowledge and technology. Alternatively, the reverse might have been true. The times of climatic variability might have been the times when large-brained and quick-witted hominids thrived relative to the canids and big cats they were competing with. This would explain why our ancestors made their exodus from Africa during the time of greatest climatic variability.

If this suggestion is correct, the 'great leap forward' in cultural complexity of the Upper Paleolithic occurred because the peoples inhabiting Europe and Asia found it possible to maintain a sufficiently large communicating population. Cultural artifacts of these people, such as the 'Venus figurines' have been found thousands of miles apart, from western Europe to southern Siberia (Saunders 1985). Shared religious beliefs and ceremonies may have been among the cultural mechanisms these people used to facilitate the sharing of cultural information between large numbers of people dispersed over a large area (Atran and Henrich 2010). Upper Paleolithic art has been interpreted as a product of religious ritual (e.g. Lewis-Williams 2002) On the other hand, Guthrie (2005) argues that the corpus of Upper Paleolithic art is remarkably naturalistic compared to the supernatural themes that are common in ethnographically well-attested contemporary hunter-gatherer art. Of course it is possible for naturalism and religion to coexist (Armstrong 2009).

The next leap forward in cultural complexity began after the start of the current geological epoch (known as 'the Holocene') about 11,000 years ago, when the climate became warmer and wetter and the extreme climatic fluctuations ceased. The last 10 millennia provide a useful natural experiment to help us choose between the cognition-lite and cognition-heavy pictures of human behaviour. If the cognition-heavy concept were correct, we would expect human behaviour to have made a rapid response to the rapid climate transition because no cultural evolution would have been necessary to allow human populations to adapt to the new environment. (Cognition-heavy theorists do not generally envision any *genetic* evolution in the Holocene either (e.g. Tooby and Cosmides 1992)). In fact, humans did *not* immediately adapt to the new, more hospitable environment. Over the millennia, populations experimented with the new ways of making a living as climate zones became relatively permanent. This slow adjustment to the new environment is predicted by a cognition-lite concept of change. Individuals don't use their innate cognition to immediately work out how best to deal with the problems and opportunities presented by new circumstances. People do have ideas and some of them are good, but new lifeways come about by many people working together over generations (Richerson and Boyd 2001). Hunter-gatherers evolved economies to exploit an array of hard-to-process plant resources. Agricultural and pastoral cultures developed, stable settlements grew and permanent trade and communication routes became established. This progressive trajectory was roughly parallel in most parts of the world albeit at rather different rates. It set off a wave of genetic evolution as our digestive physiology and ability to resist new diseases coevolved with new diets and population densities that could support epidemic diseases (Laland, Odling-Smee, and Myles 2010). This allowed a further expansion of human populations and was accompanied by further increases in cultural complexity (Richerson, Boyd, and Bettinger 2001). Large cities and associated civilizations appeared and collapsed. Christianity, Islam, Buddhism, and other systems of religious belief emerged and were disseminated widely. They provided shared beliefs and rules and unifying rituals which facilitated the spread of ideas, technology and trade among people who lived far apart and spoke different languages. Karl Jaspers (1953) classically proposed that Axial Age religions arose as the ideological part of a wave of political consolidation that was ultimately to be explained by the creation of mega-empires to defend agrarian societies from the new threat of mounted archer armies of pastoral nomads. Peter Turchin (2009) provides quantitative support for this idea.

ECONOMIC DEVELOPMENT, 'MODERNIZATION', AND 'FITNESS'

The vast changes in population size and cultural complexity triggered by economic development comprise the most recent 'leap forward' that our species has experienced. It is also the first of the 'cultural revolutions' to receive scientific scrutiny as it actually occurs. In the twentieth century Western social scientists began to focus on describing and attempting to explain the processes by which populations develop or fail to develop modern economies and the changes in attitudes and values that are associated with economic development (e.g. Baran 1957; McClelland 1961; Rostow 1960). Developing societies have been observed to abandon a set of 'traditional' values, which hinder economic growth and adopt a set of 'modern' values, which are essential for a modern economy to thrive (Inkeles and Smith 1974). Even though the values and preferences of people in modern and modernizing cultures continue to reflect aspects of the populations' cultural heritage, the empirical evidence is strong that economic development is associated with a fairly coherent cultural shift away from a set of behaviours and preferences that is shared by all pre-modern peoples towards one shared by all modern peoples (Inglehart and Baker 2000; Inglehart and Welzel 2005).

Early social theorists such as Marx, Weber, and Durkheim had personal experience of the modernization of European societies and it is not surprising that their ideas are still influential with their intellectual descendants in the West. But as the twentieth century proceeded and more populations began the process of modernization, it became possible for researchers to observe the process more objectively and outside a Western context. In *Becoming Modern*, Inkeles and Smith (1974) provided evidence of specific changes that occur in the attitudes of people in modernizing cultures. Relative to 'modern' people, members of 'traditional' cultures are disinclined to question the teachings and rulings of elders and those in authority, including the teachings and rules often characterized as 'religious'. They are more likely to feel resigned to a fate which cannot be altered by their own efforts. Their choices are limited by strong feelings of loyalty and obligation to their kin and family allies. They are more likely to believe that men and women should play different roles and that women should concentrate on children and family. They should want to marry and have a large number of children.

For scholars who have adopted a view common in Western democracies—that humans are motivated to seek freedom, comfort, pleasure and status—traditional beliefs and values inhibit the achievement of what appear to them to be 'natural' human goals (Inglehart, Foa, Peterson, and Welzel 2008; Inglehart and Welzel 2005). This idea receives some support from the literature on cognitive style (Witkin and Goodenough 1981).

Darwinian theory demands, however, that we be sceptical of the idea that humans are fundamentally motivated to seek freedom, comfort, pleasure, and status. Nor does it give us reason to believe that they will be anxious to embrace the idea that all humans have certain rights, or that everyone (including themselves and their family) should be subject to the rule of law. All living things are descended from individuals who succeeded in reproducing and have therefore inherited characteristics consistent with achieving reproductive success. We therefore do not expect animals to seek freedom, comfort, pleasure, status, and enlightenment ideals if this pursuit is inconsistent with efficiently converting resources into offspring and achieving what Darwinian theorists call 'fitness'. If one's views of animal behaviour are informed by Darwinian theory, the behaviour of people in economically developed populations is very puzzling. They exploit resources at a rate unprecedented in human history and yet the vast majority 'chooses' to limit the number of children they produce. In previous generations, the individuals who left the most descendants would have been those who increased their rate of reproduction when their circumstances were good and who favoured their kin and family allies over strangers. Evidence suggests that in traditional societies this is indeed the case (e.g. Borgerhoff-Mulder 1988; Chagnon 1988; Cronk 1989, 1991; Hill and Hurtado 1996; Irons 1979; Lindert 1985; Low 1993, 1999, 2000; Wang, Lee, and Campbell 1995).

A scientific explanation for the cultural changes of modernization has to account for why they include the abandonment of norms that promote fitness and the adoption of norms that limit their success in the production of descendants. The norms of pre-industrial societies encourage members to eschew individual desires and rebellious thoughts and concentrate on cooperating with family and friends, supported by a local community, to produce and raise the next generation (Newson and Richerson 2009). The constraints imposed by traditional norms discourage young people from wasting their time and energy in pursuits unlikely to enhance their biological fitness. For example, girls are expected to learn childcare and homemaking skills rather than the skills associated with men's work because raising a family is considered vital work. When a population begins to undergo economic development, priorities change and traditional norms begin to relax. This relaxation allows economic development to proceed. Once people feel less loyalty to their family, and are less motivated to help maintain their family, they can devote more effort to being part of the wider economy, contributing to the production of industrial goods and consuming them.

Almost all populations that have undergone economic development have experienced similar changes so it is reasonable to speculate that something about economic development causes the culture of a population to change in these ways. Newson and Richerson (2009) suggest that, in common with the previous 'revolutions' in human history, it is due to demographic change—the expansion of the communicating population. Economic development changes the structure of communities and the pattern of social interaction so that people spend less time working with and interacting with family members and family friends. Working and being educated in an industrial setting forces people to make new friends and interact with strangers. Historical accounts based on records of the literate elite can give the impression that this sort of social life was not unusual in pre-industrial times but, in fact, the vast majority of our ancestors spent most, if not all, of their lives in much smaller communities. As Bryan Wilson (1976) points out, what we call 'societies' were often no more than 'loose agglomerations of communities, partially and spasmodically drawn into association and into shared involvements by the centralized political agencies' (p. 263). Gregory King (1696) estimated that as the eighteenth century began, nearly 80% of the English population lived in villages or hamlets and few of the towns and cities, apart from London, had a population that reached 10,000.

For those of us who take a Darwinian approach to understanding human behaviour, it is particularly relevant that the social networks of our ancestors living in pre-industrial communities were dominated by their relatives and local community members they had known for most of their lives. Prior to economic development, the family is the main unit of production and the main source of support (Davis 1937/1997). Members work to acquire resources and to raise children. As children grow up, they begin to contribute and their contribution grows as they gain competence. Once they reach adulthood children may stay with their natal family, move to the family of their spouse or be employed by another family, often that of a relative. A high proportion of social interaction, therefore, is between people who see themselves as belonging to the same family or close allies of the family. Fitness-promoting norms are maintained in such a social structure because members of a family have an interest in each other's successful reproduction. Without healthy offspring the family has no future. Communication during interactions among family members is therefore likely to advise and reward behaviours that enhance family fitness (Newson et al. 2007).

With economic development, other institutions take over production (Davis 1937/1997). Travel becomes easier and young adults can move away from their family entirely to gain employment in a centre of industry or trade. Literacy becomes nearly universal, bringing urban ideas to the remotest rural communities. The resulting change is commonly described in economic terms—in terms of division of labour and increasing productivity. But as Steve Bruce (2011) says, it is not just a division of labour but a division of life; social life fragments. And, as Bryan Wilson (1976) puts it, the population becomes 'societalized'. The likely effects of these changes have been widely discussed by social scientists but they have paid little attention to a change that Darwinian theory suggests should be very important (Emlen 1995). As economic development proceeds, interaction between kin makes up a smaller and smaller proportion of people's social interaction. Even the people who stay in rural villages have increased interaction with the wider world through contact with visitors, education and the mass media. Individuals feel that they belong to a number of different social groups throughout their lives. Members of non-family groups have no particular interest in each other's fitness, so communication during the interactions among members is less likely to encourage behaviours that enhance fitness. These groups may instead have shared economic, educational or leisure interests that compete with family interests and fitness.

We have argued that it is the reduction of contact between kin and the inevitable reduction in influence from kin which causes people in developing countries to change their minds about the number of children they want to have (Newson, Postmes, Lea, and Webley 2005; Newson, Postmes, et al. 2007). The emergence of a desire to limit family size is one of the earliest changes to occur as a population begins to undergo economic development. The adoption of a 'small family norm' has been linked to the widening of social networks in every developing society that has been studied (e.g. Bongaarts and Watkins 1996; Kohler 2001; Watkins 1990; Zelinsky 1971). Its adoption is easily detected in a population. Fertility (number of children born per woman) begins to decline. Fertility is now very low or declining rapidly in almost all contemporary human populations. Some of the exceptions are theoretically quite significant. For example, Old Order Anabaptist communities, in which it is common to have families of nine children, are scattered across North America. Their populations are prosperous and fully integrated into the economies of their region. They take advantage of modern health care. But they take considerable care to isolate themselves culturally from their US and Canadian host cultures and live in small, rural communities amid their kin (Kraybill and Bowman 2001).

A large change in the size and structure of a population is unlikely to have a single contemporaneous effect. Rather, pro-family norms will decline over several generations as each succeeding generation becomes more and more modern. In the Western populations that were first to experience economic development, the adoption of the small family norm has been accompanied and followed by other cultural changes (Newson and Richerson 2009). Many of these changes further weakened the family and encouraged behaviour that reduces fitness. Even several generations after the change in the structure of communities in Western societies, cultural change is still rapid and there is no sign of it slowing down (Inglehart and Welzel 2010). It would be surprising if this ongoing cultural change did not include changes in those aspects of culture categorized as religious. A cultural evolutionary perspective is likely to shed new light on the pattern of changes that are observed.

RELIGION AND THE MODERNIZATION PROCESS

Enlightenment scholars and early social theorists observed knowledge and rational thought tackling superstition and supernaturalism. They predicted that religious belief and observance would gradually fade. Subsequent examinations of modern and modernizing societies have shown the pattern of change in religious behaviour to be very different from this (Finke and Stark 2005; Martin 2005; Norris and Inglehart 2004). European societies (including those living in colonies outside Europe) were industrializing rapidly in the nineteenth and early twentieth century and their members were adopting modern values. Birth-rates were plummeting, causing amazement and consternation among politicians and scholars (e.g. National Birth-Rate Commission 1916). Yet many populations were experiencing religious revivals at this time (McLoughlin 1978). It was a time of much religious discussion, innovation and church-building. As the twentieth century progressed, however, these societies became more secular.

As the influence of organized religion waned in Western societies, many of the non-European populations that began to undergo economic development during the middle part of the twentieth century began to experience religious revivals, often described as 'fundamentalist movements'. Such movements are found today in Africa, the Middle East, Latin America, and Asia and include practitioners of Judaism, Hinduism, Islam, Sikhism, neo-Confucianism, Christianity, and Buddhism (Emerson and Hartman 2006). Their popularity has inspired a number of commentators to describe a 'clash of civilizations' between their own secular Western societies and populations of faith, and particularly of the Islamic faith, which refuse to embrace rationality, tolerance, and other modern values (e.g. Harris 2004; Huntington 1993).

Norris and Inglehart (2004) suggest that those promoting secularism may have reason for more optimism. Using data from the European and World Values Survey (2009), they compared the religious beliefs and practices of over a quarter of a million people in more than 70 countries who were surveyed between 1981 and 2001. Participants in the surveys included members of all major religions, although Christians made up a much larger proportion of the surveyed population than they do of the world population. Their analysis of the data shows that 'the publics of virtually all advanced industrialized societies have been moving toward more secular orientation (pp. 24–5)'. But they attribute this trend not to increasing rationality but to rising levels of security. They argue that their data disputes the suggestion that the 'demand' for religion is constant (Iannaccone, Finke, and Stark 1997) because religious belief and observance increase with stress and insecurity. The influence of religious institutions weakens with modernization, they say, because rising levels of education and wealth makes people's lives less stressful. If this interpretation of the data is correct, then people in developing countries will start behaving more like Westerners if they enjoy increased security and other advantages of economic development that Westerners enjoy. Norris and Inglehart warn, however, that poor people are having more children and they attribute this to their religious values, which emphasize a traditional role for women and limit their lives to producing and raising as many children as possible. Having so many children will reduce their chances of becoming wealthier and more secure. They are likely, therefore, to continue to eschew secularism and because of their rapid population growth, religious believers will make up a growing proportion of the world's population.

A cultural evolutionary analysis of the European and World Value Survey and demographic trend data suggests very different explanations for the cross-national differences in religiosity. First of all, as explained above, from a Darwinian perspective, the 'traditional family values' of pre-industrialized societies are explained by the high level of social interaction between kin that occurs in pre-modern populations. The idea that the abandonment of these values is prevented or slowed by the influence of religion is not supported. For example, as Figure 11.1 shows, in Iran the adoption of family limitation occurred after the 1979 Islamic revolution and once it began it proceeded at the same pace as it had in China. Now both Iran and China have lower fertility than the United States and a number of European countries. The adoption of family limitation is occurring more slowly in India, which has a more heterogeneous population. What may be called 'traditional family value' norms

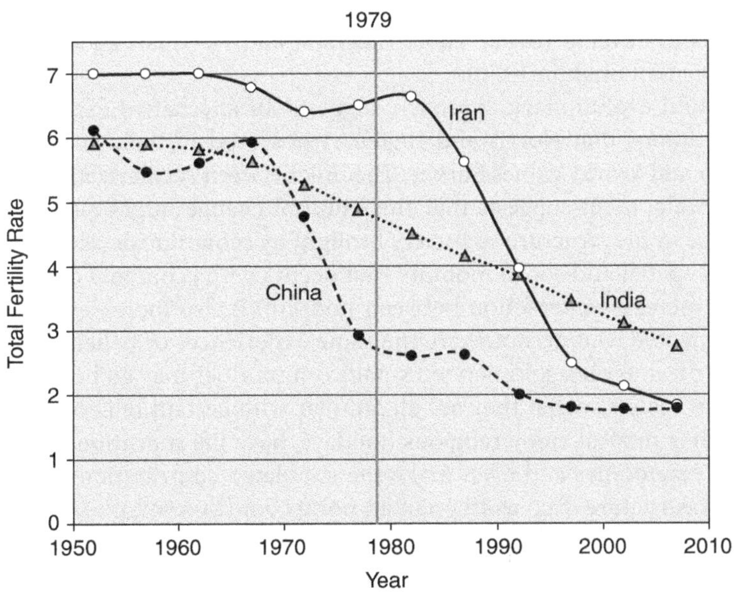

Fig. 11.1. Total fertility rate, the number of children produced per woman per year, has declined in all countries which are experiencing economic development. Fertility began to decline in Iran after it became an Islamic republic. The decline began earlier in China than in Iran but they proceeded at the same pace even though religious belief and observance is higher in Iran. The small-family norm is spreading more slowly in the large and very diverse population of India. 1979 was the year of the Islamist revolution in Iran and also the year that China instituted its 'one child policy'.

begin to relax once economic development allows communities to be less kin-based. But the relaxation is part of a *process* of change. People don't change their minds instantaneously; the process takes generations to play out. The adoption of family limitation occurs relatively early in the process. Changes in other norms, such as those prescribing gender roles and sexual conduct, have tended to occur later (Newson 2009). Members of a modernizing population who continue to hold more traditional norms about gender and sexual conduct may categorize these values among their religious beliefs and they may use religious texts to justify them, but this does not mean that these texts are the source of these values.

Norris and Inglehart's suggestion that higher levels of insecurity lead to increased demand for religion may be supported by an analysis comparing countries that began to modernize some time ago with those that began to modernize more recently. But it ignores the pattern of religious change observed at early stages of economic development. When a population begins to undergo economic development, employment and education opportunities increase. People become more prosperous and secure. Yet, in many populations, religious belief and observance remain high or even increase as the society begins to develop (Bruce 2000; Emerson and Hartman 2006; Finke and Stark 2005; McLoughlin 1978).

A cultural evolutionary approach suggests an alternative explanation for the associations that Norris and Inglehart observed in their analysis of the European and World Values Survey. The link between secularization and economic development suggests that this cultural change might also be due to the change in the structure of society brought by economic development. The increased spatial and social mobility that occurs with economic development not only increases interaction between non-kin, it also increases interaction between people who do not share the same experiences or beliefs. A population living in loosely agglomerated small communities may all be of the same religion in the sense that they are all familiar with certain religious texts and rituals. They may all enjoy religious holidays, have life transitions marked by religious ceremonies and even make the stipulated contributions to the religious infrastructure. But, as Pyysiäinen points out (this volume), some forms of religion are more convincing and more easily remembered than others, and this is bound to result in regional differences in the details of customs, religious practices, and superstitions. Clusters of villages and hamlets whose inhabitants regularly meet with one another and frequently intermarry are likely to share basically the same beliefs, but when such communities are isolated and have little contact with the religious elite, a wide range of heretical beliefs and practices are likely to emerge. The historical record shows that this was indeed the case in Europe (Foster 1960; Thomas 1971; Watkins 1991).

Economic development provides people from isolated communities with the means and opportunity to travel to centres of industry and commerce. Upon arrival they become part of a community made up of people from different locales whose beliefs, values, customs, and dialects are different. Newcomers find that some of their beliefs are considered bizarre by many of their new neighbours. And these neighbours have incredible beliefs and strange customs of their own. Back home in the village, people shared a small number of limited experiences and group memories. Many of their beliefs had a 'social validity' (Turner 1991) and were seldom questioned. On moving to a larger community, especially a diverse immigrant community, such villagers find these beliefs questioned. Life in the village doesn't prepare them for this. They lack a basis on which to decide which beliefs might be valid and which to abandon. Urban areas with jobs that attract people from country villages receive a steady stream of lonely, awkward and confused young immigrants. In such an environment, institutions that provide guidance such as political movements, learned societies, and religious congregations can be very helpful and very influential. Carvalho (2013) provides an interesting analysis of veiling, which becomes more common in Muslim populations as they begin the early stages of modernization. He suggests that veiling permits young women to make a visible statement of their commitment to their still-traditional community while at the same time seeking modern educations and employment in the modern economy. In general, fundamentalist religious movements are products of the interaction of urbanizing traditional communities with modernism (Al Sayyad and Massoumi 2011). They are one part adaptation to modernity and one part resistance to it. It is not surprising that among the more successful religions that emerge in such environments are those which base the validity of their belief on something with which potential members are familiar, such as a sacred text. Among the five ideological characteristics of fundamentalist groups, identified by the Fundamentalisms Project (Almond, Appleby, and Sivan 2003) is a belief that the text of the religious tradition is 'of divine (inspired) origin and true and accurate in all particulars'. In religions like Hinduism, which do not have a single clear sacred text, fundamentalist movements are inclined to nominate one text or a set of texts to this position. Thus, we agree with Norris and Inglehart (2004) that religious belief helps to relieve insecurity but the insecurity is not primarily caused by poverty. It is mainly due to the lack of a clear social identity as well as the lack of functional social networks to give support and advice on such things as finding a spouse or medical help.

It may be that the cultural changes of modernization progress towards more secularism and greater tolerance of religious differences because the children and grandchildren of the immigrants from the village grow up surrounded by diversity. They have the chance and the necessity to develop more sophisticated heuristics to allow them to choose among beliefs and also to choose not to choose—to decide that some things simply cannot be known for certain. They feel secure in a community with people who hold different beliefs. Thus, our hypothesis is that the influence of religious organizations tends to rise after the start of economic development and then begins to decline as modernization proceeds. But these cultural changes are not instantaneous and, like the changes in reproductive norms, they can take generations to play out.

If this cultural evolutionary interpretation of secularization is correct, then the differences in religiosity observed in contemporary populations may not be the result of deep-seated cultural differences between the populations. The populations may actually be experiencing similar cultural evolutionary processes, and the countries which began economic development earlier may simply be further along that process. Modernization includes a large number of highly correlated changes—cultural changes and changes in lifestyle. The suggestion that one of the changes causes another—that poverty causes religiosity and religiosity causes high birth-rates, for example—is consistent with these observations. But correlation does not imply causation. The same correlations would be observed if the changes were part of an evolutionary process that began with the profound change in the structure of societies that occurred early in economic development before many of the changes were measurable.

The European World Values Survey (2009) data analyzed by Norris and Inglehart (2004) can also be analyzed from a cultural evolutionary perspective. If countries are experiencing a cultural evolutionary process and fertility begins early on in that process, then the year fertility begins to decline in a country provides a culturally neutral index of how far along the process the different countries are. We found that this single variable, the year fertility begins to decline in a country, explains about 50% of the variance between countries in mean response to questions about religious attendance (Figure 11.2) and belief in hell (Figure 11.3). It explains about a quarter of the variance in belief in God and life after death. Like Norris and Inglehart, we found the United States to be a significant outlier. A number of explanations have been offered for the high proportion of Americans for whom religion is important (Wald and Calhoun-Brown 2007), such as the entrepreneurial nature of religion in the United States (Iannaccone et al. 1997) and the high levels of inequality (Norris and Inglehart 2004).

An alternative explanation, consistent with our cultural evolutionary approach, is that important subpopulations in the US have entered the modernization processes later than others. For example, the south-eastern US, where many Conservative Protestants have their roots, lagged the rest of the country in urbanization in the early twentieth century, continuing to live in small communities (Fischer and Hout 2006). Also, in 1965, changes to legislation governing immigration to the United States increased rates of immigration and dramatically decreased the proportion of immigrants coming from Europe (Segal, Elliott, and Mayadas 2010). In 2007, 12.6% of Americans were foreign-born and approximately 85% of immigrants were from less-developed countries like Mexico and the Philippines, where fertility only began to decline in the latter part of the twentieth century. Thus, even though much of the United

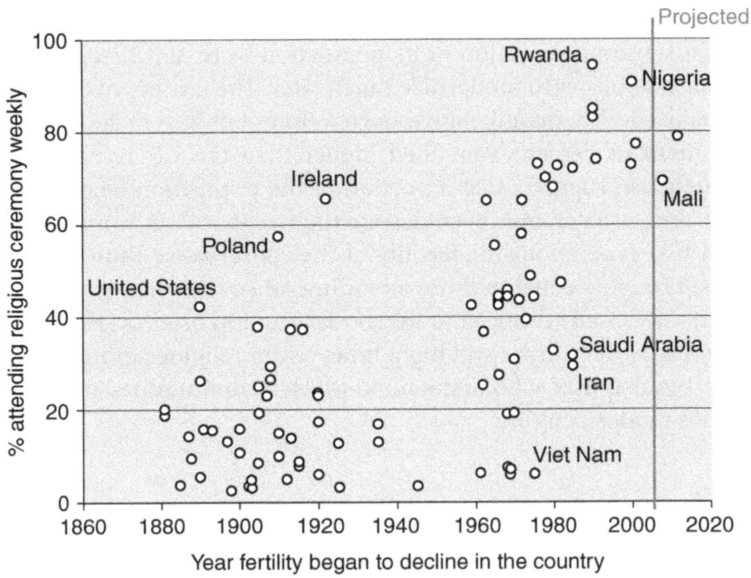

Fig. 11.2. Percentage of population in 87 countries saying they attended a religious ceremony once a week or more*

*Based on responses to a European and World Values Survey (2009) questionnaire issued to a representative sample of the countries' population between 1994 and 2008.

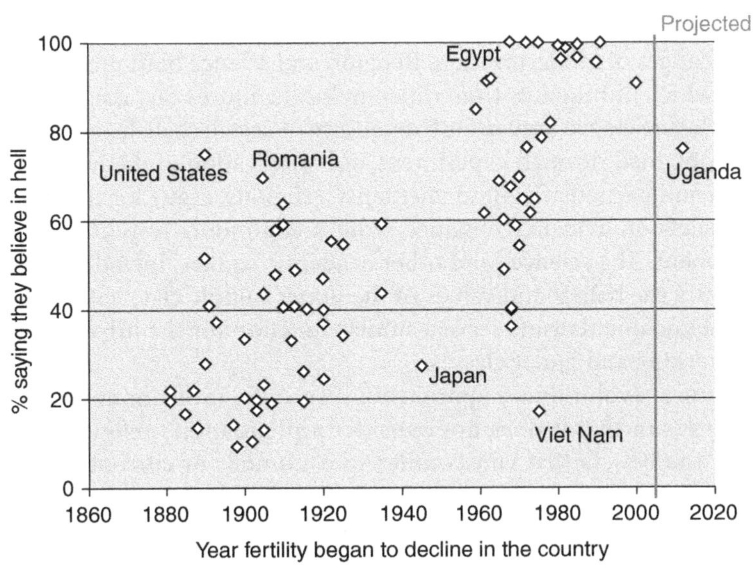

Fig. 11.3. Percentage of population in 77 countries saying they believe in hell*

*Based on responses to a European and World Values Survey (2009) questionnaire issued to a representative sample of the countries' population between 1994 and 2008.

States began the modernization process at the same time as West European countries, a substantial portion of its population were, until recently, part of populations that began to modernize much later. The fertility of Conservative Protestants in the US, mainly native-born Whites but also including many of African-American descent, remained higher than the US average until the 1970s, which also suggests that a portion of the population began the modernization process later. However, despite their espousal of 'family values', for more than two generations the fertility of the conservative Protestants in the US has been roughly equal to those attending other churches, which suggests that they are now well advanced in the modernization process (Hout, Greeley, and Wilde 2001). Fertility is still high, however, in religious groups known as 'Old Order Anabaptists' who maintain kin-based communities and have little contact with modern culture.

CONCLUSION

Our cultural evolutionary explanation of religious change and modernization differs in several ways from the narrative of what David Martin (2005, p. 8) calls the variety of secularization stories. We do not assume that 'Once there was *a* religious past, and that this has bit by bit emerged as *the* secular future' and we do not suggest that 'The natural (often equated with science) wins out over religion'. In fact, we see religion as playing a similar role to science in the cultural changes of modernization. Religion and science both provide mechanisms by which immigrants from different local cultures can assign validity to certain beliefs. Science argues for the validity of certain beliefs on the basis of evidence obtained through experiment and observation of the natural world. Religions, and particularly 'fundamentalist' religions, argue for the validity of certain beliefs on evidence obtained from a commonly respected source of enlightenment. The sciences and other academic studies, including theology, help to unify the beliefs and values of the upper middle class elite of modern societies. Fundamentalisms serve a similar function for the urbanizing lower middle, working and under classes.

The cultural evolutionary approach is also different from many secularization stories in that it does not consider a population's religion to be the source of the beliefs that run counter to a number of currently emerging Western beliefs about sexual behaviour. People who are most influenced by religion may also hold more 'traditional' beliefs about sexuality. But a Darwinian approach suggests that beliefs about sexuality, marriage, sexual division of labour (as well as nepotism and respect for elders) are a product, not of religion, but of the kin-based communities in which most people were socialized prior to economic development. Religious texts may include references to such beliefs but these texts are also often the products of kin-based communities. Religions can encourage their members to be tolerant and to respect human rights but the beliefs that support the tolerance and respect have to evolve in the population and cannot be imposed from 'above' or outside. This is consistent with the observations of some scholars studying the adoption of beliefs about human rights by Islamic populations (e.g. Dalacoura 2007).

Finally, we support our suggestion that the cultural evolutionary approach is 'scientific' by making a prediction. We predict that cultural differences between the West and many developing countries will be reconciled peacefully despite religious differences. The cultural evolutionary process known as modernization is proceeding in both cultural groups. Immigration and communication technologies are strengthening social links. Social links between the groups will help the cultures of populations to converge. The near future provides a test for this optimistic prediction. However, 'modernization' is a continuing process. What is 'modern' will keep changing. Indeed, modern behaviour in its current form is clearly unsustainable. Not only will it not be possible to keep up the levels of consumption which many modern and modernizing people aspire to, but we will simply not be able to maintain our populations. In many modern societies, fertility is below replacement. Selection even on cultural variation will favour subpopulations that maintain above-replacement fertility. In the long run, perhaps the population of the world will be dominated by Anabaptists, Ultra-orthodox Jews, Roma, and whatever other cultures manage to resist modernity. Perhaps the meek shall inherit the earth.

REFERENCES

Al Sayyad, N. and M. Massoumi. 2011. *The Fundamentalist City: Religiosity and the Remaking of Urban Space*. Oxford: Routledge.

Alchian, A. A. 1950. Uncertainty, evolution and economic theory. *Journal of Political Economy 58*: 211–22.

Almond, G. A., R. S. Appleby, and E. Sivan. 2003. *Strong Religion: The Rise of Fundamentalisms Around the World*. Chicago, IL: Chicago University Press.

Armstrong, K. 2009. *The Case for God*. New York: Alfred A. Knopf.

Atkinson, Q. D., R. D. Gray, and A. J. Drummond. 2008. mtDNA variation predicts population size in humans and reveals a major southern Asian chapter in human prehistory. *Molecular Biology and Evolution 25*(2): 468–74.

Atran, S. and J. Henrich. 2010. The evolution of religion: How cognitive by-products, adaptive learning heuristics, and group competition generate deep commitments to prosocial religions. *Biological Theory 5*(1): 18–30.

Baran, P. A. 1957. *The Political Economy of Growth*. New York: Monthly Review Press.

Bobe, R. and M. G. Leakey. 2009. Ecology of plio-pleistocene mammals in the Omo-Turkana basin and the emergence of Homo. In *The First Humans: Origins and*

Early Evolution of the Genus Homo. Eds. F. E. Grine, J. G. Fleagle, and R. E. Leakey. New York: Springer, pp. 173–84.

Bongaarts, J. and S. C. Watkins. 1996. Social interactions and contemporary fertility transitions. *Population And Development Review 22*(4): 639–82.

Borgerhoff-Mulder, M. 1988. Behavioural ecology in traditional societies. *TREE 3*(10): 260–4.

Bouckaert, R., P. Lemey, M. Dunn, S. J. Greenhill, A. V. Alekseyenko, A. Drummond, and Q. D. Atkinson. 2012. Mapping the origins and expansion of the Indo-European language family. *Science 337*(6097): 957–60.

Boyd, R. and P. J. Richerson. 1985. *Culture and the Evolutionary Process*. Chicago: University of Chicago Press.

——. 1992. How microevolutionary processes give rise to history. In *History and Evolution*. Eds. M. H. Nitecki and D. V. Nitecki, 179–209. Albany: State University of New York Press.

Boyd, R., P. J. Richerson, and J. Henrich. 2011. The cultural niche. *Proceedings of the National Academy of Sciences of the United States of America 108*: 10918–25.

Boyer, P. 2001. *Religion Explained: The Evolutionary Origins of Religious Thought*. New York: Basic Books.

Bruce, S. 2000. *Fundamentalism*. Malden, MA: Blackwell.

——. 2011. *Secularization: In Defence of an Unfashionable Theory*. Oxford: Oxford University Press.

Campbell, D. T. 1960. Blind variation and selective retention in creative thought as in other knowledge processes. *Psychological Review 67*: 380–400.

Carey, S. 2009. *The Origin of Concepts*. New York: Oxford University Press.

Carvalho, J. P. 2013. Veiling. *Quarterly Journal of Economics 128*(1): 337–70.

Cavalli-Sforza, L. L. and M. W. Feldman. 1973. Models for cultural inheritance: I. Group mean and within group variation. *Theoretical Population Biology 4*: 42–55.

——. 1981. *Cultural Transmission and Evolution: A Quantitative Approach*. Princeton, NJ: Princeton University Press.

Chagnon, N. A. 1988. Life histories, blood revenge, and warfare in a tribal population. *Science 239*: 985–92.

Cronk, L. 1989. Low socio-economic status and female biased parental investment: The Mukogodo example. *Amercian Anthropologist 91*: 441–29.

——. 1991. Human behavioral ecology. *Annual Review of Anthropology 20*: 25–53.

Dalacoura, K. 2007. *Islam, Liberalism and Human Rights: Implications for International Relations*. London: I.B. Tauris.

Darwin, C. 1859. *On the Origin of Species by Means of Natural Selection, or, The Preservation of Favoured Races in the Struggle for Life*. London: John Murray.

——. 1874. *The Descent of Man and Selection in Relation to Sex* (2nd edn). New York: American Home Library.

Davis, K. 1937/1997. Kingsley Davis on reproductive institutions and the pressure for population. *Population and Development Review 23*(3): 611–24.

Diamond, J. 1978. The Tasmanians: the longest isolation, the simplest technology. *Nature 273*: 185–6.

——. 1997. *Guns, Germs, and Steel: The Fates of Human Societies*: Norton; Jonathan Cape/Random House.

Emerson, M. O. and D. Hartman. 2006. The rise of religious fundamentalism. *Annual Review of Sociology 32*: 127–44.

Emlen, S. T. 1995. An evolutionary theory of the family. *Proceedings of the National Academy of Sciences of the United States of America 92*: 8092–9.

European and World Values Surveys. 2009. Five-wave integrated Data File 1981-2007 v20090914. Retrieved 19 Nov from European Values Study Foundation and World Values Survey Association. Available online at <http://www.worldvaluessurvey.org/>

Finke, R. and R. Stark. 2005. *The Churching of America, 1776–2005: Winners and Losers in Our Religious Economy*. New Brunswick, NJ: Rutgers University Press.

Fischer, C. S. and M. Hout. 2006. *Century of Difference*. New York: Russell Sage Foundation.

Foster, G. M. 1960. *Culture and Conquest: America's Spanish Heritage*. New York: Wenner-Gren Foundation for Anthropological Research.

Gerard, R. W., C. Kluckhohn, and A. Rapoport. 1956. Biological and cultural evolutions. *Behavioral Sci. 1*: 6–34.

Gervais, W. M., A. K. Willard, A. Norenzayan, and J. Henrich. 2011. The cultural transmission of faith: Why innate intuitions are necessary, but insufficient, to explain religious belief. *Religion 41*: 389–410.

Guthrie, R. D. 2005. *The Nature of Paleolithic Art*. Chicago: Chicago University Press.

Harris, L. 2004. *Civilization and its Enemies*. New York: The Free Press.

Harris, P. L. and M. A. Koenig. 2006. Trust in testimony: How children learn about science and religion. *Child Development 77*(3): 505–24.

Harris, P. L., E. S. Pasquini, S. Duke, J. J. Asscher, and F. Pons. 2006. Germs and angels: The role of testimony in young children's ontology. *Developmental Science 9*(1): 76–96.

Hauser, M. D., N. Chomsky, and W. T. Fitch. 2002. The faculty of language: What is it, who has it, and how did it evolve? *Science 298*(5598): 1569–79.

Henrich, J. 2004. Demography and cultural evolution: Why adaptive cultural processes produced maladaptive losses in Tasmania. *American Antiquity 69*(2): 197–214.

Henrich, J. and N. Henrich. 2010. The evolution of cultural adaptations: Fijian food taboos protect against dangerous marine toxins. *Proceedings of the Royal Society B: Biological Sciences*. doi: 10.1098/rspb.2010.1191.

Hill, K. and A. M. Hurtado. 1996. *Ache Life History: The Ecology and Demography of a Foraging People*. New York: Aldine de Gruyter.

Hodgson, G. M. 2004. *The Evolution of Institutional Economics: Agency, Structure and Darwinism in American Institutionalism*. London: Routledge.

Hout, M., A. M. Greeley, and M. J. Wilde. 2001. The demographic imperative in religious change in the United States. *American Journal of Sociology 107*(2): 468–86.

Huntington, S. P. 1993. The clash of civilizations. *Foreign Affairs 72*(3): 22–49.

Hurford, J. R. 2011. *The Origins of Grammar: Language in the Light of Evolution II*. New York: Oxford University Press.

Iannaccone, L. R., R. Finke, and R. Stark. 1997. Deregulating religion: The economics of church and state. *Economic Enquiry 35*: 350–64.

Inglehart, R. and W. E. Baker. 2000. Modernization, cultural change, and the persistence of traditional values. *American Sociological Review 65*(1): 19–51.

Inglehart, R., R. Foa, C. Peterson, and C. Welzel. 2008. Development, freedom, and rising happiness: A global perspective (1981–2007). *Perspectives on Psychological Science 3*. 264–85. doi: 10.1111/j.1745-6924.2008.00078.x

Inglehart, R. and C. Welzel. 2005. *Modernization, Cultural Change, and Democracy: The Human Development Sequence*. New York: Cambridge University Press.

——. 2010. Changing mass priorities: The link between modernization and democracy. *Perspectives on Politics 8*(2): 551–67.

Inkeles, A. and D. H. Smith. 1974. *Becoming Modern: Individual Change in Six Developing Countries*. Cambridge, MA: Harvard University Press.

Irons, W. 1979. Cultural and biological success. In *Evolutionary Biology and Human Social Behavior*. Eds. N. A. Chagnon and W. Irons. North Scituate, MA: Duxbury Press. 257–72.

Jacobs, Z., R. G. Roberts, R. F. Galbraith, H. J. Deacon, R. Grun, A. Mackay and L. Wadley. 2008. Ages for the Middle Stone Age of Southern Africa: Implications for human behavior and dispersal. *Science 322*: 733–5.

Jaspers, K. 1953. *The Origin and Goal of History*. New York: Routledge and Kegan Paul.

King, G. 1696. *Natural and Political Observations and Conclusions Upon the State and Condition of England 1696*. Gregory King Collection Ms. Hut 3, Special Collections, Milton S. Eisenhower Library, Johns Hopkins University.

Klein, R. G. 2009. *The Human Career: Human Biological and Cultural Origins, 3rd edn*. Chicago: Chicago University Press.

Kline, M. A. and R. Boyd. 2010. Population size predicts technological complexity in Oceania. *Proceedings of the Royal Society B 277*: 2559–64.

Kohler, H. P. 2001. *Fertility and Social Interactions*. Oxford: Oxford University Press.

Kraybill, D. B. and C. F. Bowman. 2001. *On the Backroad to Heaven: Old Order Hutterites, Mennonites, Amish, and Brethren*. Baltimore: Johns Hopkins University Press.

Laland, K. N., J. Odling-Smee, and S. Myles. 2010. How culture shaped the human genome: bringing genetics and the human sciences together. *Nature Reviews Genetics 11*: 137–48.

Lewis-Williams, D. 2002. *The Mind in the Cave: Consciousness and the Origins of Art*. London: Thames and Hudson.

Lindert, P. H. 1985. English population, prices, and wages, 1541–1913. *Journal of Interdisciplinary History 15*: 609–34.

Loftus, E. F. 1996. *Eyewitness Testimony*. Cambridge, MA: Harvard University Press.

Low, B. S. 1993. Ecological demography: A synthetic focus in evolutionary anthropology. *Evolutionary Anthropology 1*: 177–87.

——. 1999. Sex, wealth and fertility: Old rules in new environments. In *Human Behavior and Adaptations: An Anthropological Perspective*. Eds. L. Cronk, N. A. Chagnon, and W. Irons. New York: Aldine de Gruyter, 323–44.

——. 2000. *Why Sex Matters*. Princeton, N.J.: Princeton University Press.

Martin, D. 2005. *On Secularization: Towards a Revised General Theory*. Aldershot, Hants: Ashgate Publishing Limited.

Martrat, B., J. O. Grimalt, N. J. Shackleton, L. de Abreu, M. A. Hutterli, and T. F. Stocker. 2007. Four climate cycles of recurring deep and surface water destabilizations on the Iberian margin. *Science 317*: 502–7.

McClelland, D. 1961. *The Achieving Society*. Princeton, N.J.: D. Van Nostrand.
McElreath, R., R. Boyd, and P. J. Richerson. 2003. Shared norms and the evolution of ethnic markers. *Current Anthropology 44*: 122–9.
McLoughlin, W. G. 1978. *Revivals, Awakenings, and Reform: An Essay on Religion and Social Change in America. 1607–1977*. Chicago: Chicago University Press.
Mesoudi, A. 2007. Using the methods of experimental social psychology to study cultural evolution. *Journal of Social, Evolutionary, and Cultural Psychology 1*(2): 35–58.
Mesoudi, A., A. Whiten, and K. N. Laland. 2004. Is human cultural evolution Darwinian? Evidence reviewed from the perspective of 'The Origin of Species'. *Evolution 58*(1): 1–11.
National Birth-Rate Commission. 1916. *The Declining Birth-Rate: Its Causes and Effects*. London: Chapman and Hall.
Newson, L. 2009. Cultural versus reproductive success: Why does economic development bring new tradeoffs? *American Journal of Human Biology 21*(4): 464–71.
Newson, L., T. Postmes, S. E. G. Lea, and P. Webley. 2005. Why are modern families small? Toward an evolutionary and cultural explanation for the demographic transition. *Personality And Social Psychology Review 9*(4): 360–75.
Newson, L., T. Postmes, S. E. G. Lea, P. Webley, P. J. Richerson, and R. McElreath. 2007. Influences on Communication about Reproduction: The Cultural Evolution of Low Fertility. *Evolution and Human Behavior 28*: 199–210.
Newson, L. and P. J. Richerson. 2009. Why do people become modern?: A Darwinian Mechanism. *Population and Development Review 35*(1): 117–58.
Newson, L., P. J. Richerson, and R. Boyd. 2007. Cultural evolution and the shaping of cultural diversity. In *Handbook of Cultural Diversity*. Eds. S. Kitayama and D. Cohen. New York: Guilford Press, 454–76.
Norris, P. and R. Inglehart. 2004. *Sacred and Secular: Religion and Politics Worldwide*. Cambridge: Cambridge University Press.
Perreault, C. 2012. The pace of cultural evolution. *PLoS ONE, 7*, e45150. doi: doi:10.1371.
Powell, A., S. Shennan, and M. Thomas. 2009. Late Pleistocene demography and the appearance of modern human behavior. *Science 324*: 1298–301.
Richards, R. J. 1987. *Darwin and the Emergence of Evolutionary Theories of Mind and Behavior*. Chicago: University of Chicago Press.
Richerson, P. J. and R. Boyd. 1987. Simple models of complex phenomena: The case of cultural evolution. In *The Latest on the Best: Essays on Evolution and Optimality*. Ed. J. Dupré. Cambridge: MIT Press, 27–52.
——. 2001. Institutional evolution in the holocene: The rise of complex societies. In *The Origin of Human Social Institutions (Vol. 110)*. Ed. W. G. Runciman. Oxford: Oxford University Press, 197–234.
——. 2005. *Not By Genes Alone: How Culture Transformed Human Evolution*. Chicago: Chicago University Press.
——. 2010. The Darwinian theory of human cultural evolution and gene-culture coevolution. In *Evolution Since Darwin: The First 150 Years*. Eds. M. A. Bell, D. J. Futuyma, W. F. Eanes, and J. S. Levinton. Sunderland, MA: Sinauer. Available online at <http://ssrn.com/abstract=1616311>.
Richerson, P. J., R. Boyd, and R. Bettinger. 2001. Was agriculture impossible during the Pleistocene but mandatory during the Holocene? A climate change hypothesis. *American Antiquity 66*(3): 387–411.

Rostow, W. W. 1960. *The Stages of Economic Growth: A Non-communist Manifesto*. Cambridge: Cambridge University Press.

Saunders, N. K. 1985. *Prehistoric Art in Europe (2nd edn)*. New Haven, CT: Yale University Press.

Segal, U. A., D. Elliott, and N. S. Mayadas. 2010. *Immigration Worldwide: Policies, Practices, and Trends*. Oxford: Oxford University Press.

Shea, J. J. 2008. The Middle Stone Age archaeology of the Lower Omo Valley Kibish Formation: Excavations, lithic assemblages, and inferred patterns of early Homo sapiens behavior. *Journal of Human Evolution 55*(3): 448–85.

Slingerland, E. 2008. *What Science Offers the Humanities: Integrating Body and Mind*. New York: Cambridge University Press.

Thomas, K. 1971. *Religion and the Decline of Magic*. Oxford: Weidenfeld and Nicholson.

Tomasello, M. 1999. *The Cultural Origins of Human Cognition*. Cambridge, MA: Harvard University Press.

Tooby, J. and L. Cosmides. 1992. The psychological foundations of culture. In *The Adapted Mind: Evolutionary Psychology and the Generation of Culture*. Eds. J. Barkow, L. Cosmides, and J. Tooby. New York: Oxford University Press, 19–136.

Turchin, P. 2009. A theory for the formation of large empires. *Journal of Global History*, 4: 191–217.

Turner, J. C. 1991. *Social Influence*. Milton Keynes: Open University Press.

Wald, K. D. and A. Calhoun-Brown. 2007. *Religion and Politics in the United States*. Lanham, Maryland: Rowman and Littlefield.

Wang, F., J. Lee, and C. Campbell. 1995. Marital fertility control among the Qing nobility: Implications for two types of preventive checks. *Population Studies 49*: 383–400.

Watkins, S. C. 1990. From local to national communities—The transformation of demographic regimes in Western-Europe, 1870-1960. *Population and Development Review 16*(2): 241–72.

——. 1991. *From Provinces into Nations: Demographic Integration in Western Europe, 1870-1960*. Princeton, NJ: Princeton University Press.

White, T. D., B. Asfaw, D. DeGusta, H. Gilbert, G. D. Richards, G. Suwa, and F. C. Howell. 2003. Pleistocene Homo sapiens from Middle Awash, Ethiopia. *Nature 423*: 742–47.

Whiten, A., R. A. Hinde, K. N. Laland, and C. B. Stringer. 2011. Culture evolves. *Philosophical Transactions of the Royal Society (B) 366*: 938–48.

Wilson, B. 1976. Aspects of secularization in the West. *Japanese Journal of Religious Studies 3*(4): 259–76.

Witkin, H. A. and D. R. Goodenough. 1981. *Cognitive Styles, Essence and Origins: Field Dependence and Field Independence*. New York: International Universities Press.

Zelinsky, W. 1971. The hypothesis of the mobility transition. *The Geographical Review 61*: 219–49.

12

Artificial Intelligence Models of Religious Evolution

WILLIAM SIMS BAINBRIDGE

INTRODUCTION

Sociologists tend to believe that the form of evolution that has dominated the past few thousand years, during which religious institutions emerged, was cultural rather than biological, although biological evolution set the conditions on which cultural evolution could build. Like Newson and Richerson (this volume) my concern in this chapter is to explore such a cultural evolutionary approach. I mean to explain how studies of artificial intelligence, one of the founding disciplines within cognitive science, can converge with sociology to model the forms of social cognition responsible for the emergence and evolution of religion. The fundamental assumptions behind computer simulations of faith tend to conceptualize it as a form of cognitive error, perhaps induced by unmet needs and social influences, if not innate to the evolution-determined structure of the human mind. Artificial social intelligence studies of religion, however, tend to focus on social networks, or on the notions of goal-directed human action communicated through dynamic social interactions.

If artificial intelligence is to realize its potential to enhance our understanding of human religion, several difficult challenges must be overcome, including: (1) assembling insights from many different schools of thought and topic areas, (2) developing new technical means for data analysis and theory development, (3) overcoming resistance from well-meaning but technophobic scholars in religious studies, and (4) achieving a perspective that rises above any particular religious faith. Cognitive science can contribute to scientific research on religion in many ways, but achievement of these goals would especially enhance our understanding of how human hopes have shaped the evolution of society's institutions and the conceptual world inhabited by each individual person (Bainbridge 1985).

THE CHALLENGES

The cover of each issue of a leading publication, *Cognitive Science: A Multidisciplinary Journal* lists the names of seven disciplines that came together to create the new field: artificial intelligence, education, linguistics, neuroscience, philosophy, psychology, and anthropology. Notable for their absence are two highly cognitive social sciences, sociology and political science, and it is left ambiguous whether philosophy includes theology. The inclusion of anthropology meant that cognitive science could study the religious expressions of pre-industrial peoples (e.g. Whitehouse 1995), but the exclusion of sociology left out the main social science that studies religion in modern societies.

Opponents of religion, including some New Atheist intellectuals who draw upon cognitive science in their critique of faith, seem to feel that the existence of God is the main issue (Dennett 1995; Dawkins 2006; cf. Bainbridge 2009, 2010). But over the full historical sweep of human cultural evolution, cognitions concerning a range of other supernatural ideas have been equally important, such as: competition among multiple gods that provides an explanation of the maelstrom of forces that constitute the natural world, veneration of deceased ancestors who remain somewhat real in our memories even after their deaths, rhetorical tactics intended to impose a moral order upon society that is more than simply the net sum of individual self-interests, the harnessing of magical techniques to achieve personal goals that otherwise are beyond reach, and elevating one's self-perception to achieve a sense of unity and a hope of immortality.

The term *artificial intelligence* (AI) can be defined in many ways, and it has often been a source of misunderstanding. AI was launched in a small conference at Dartmouth College in 1956, with exceedingly ambitious goals to give machines all the cognitive abilities of human beings, inspired by faith that very rapid progress was possible (McCorduck 2004; Bainbridge 2012). Some progress was achieved in the following decades, but slowly, and one consequence was a certain schizophrenia around AI. Some proponents made exaggerated claims for what their own particular methods could accomplish, while critics dismissed AI as a vacuous fad. Within computer science today, the term *artificial intelligence* is often avoided, because the general public takes it to mean machine emulation of the human mind, while recent technical work focuses much more on finding algorithms or data structures to solve specific difficult problems, often ones the human mind cannot easily solve.

A leading textbook titled *Artificial Intelligence* identifies six subfields: natural language processing, knowledge representation, automated reasoning, machine learning, computer vision and robotics (Russell and Norvig 2003, pp. 2–3). Some of these, notably natural language processing and machine learning, can be used to analyze empirical data, such as religious texts and the structure of social relations within religious movements. Knowledge representation, automated reasoning, and machine learning are especially useful for computer simulations of religious behaviour, designed to improve the conceptual rigour of theories such as those about religious conversion through social networks of communication.

A classic example of AI research relevant to religion is *The Evolution of Cooperation* by political scientist Robert Axelrod (1984). While much recent AI work uses highly complex computer programs based on sophisticated algorithms, Axelrod's work achieved a great deal on the basis of a very simple program. Initial results were reported through an article published in the journal *Science*, written in collaboration with evolutionary biologist William Hamilton, and the research was solidly based in evolutionary theory, while taking advantage of computer simulation techniques (Axelrod and Hamilton 1981). For his book, Axelrod solicited strategies from colleagues that could be programmed into computer-simulated agents who would play the 'prisoner's dilemma' game in pairs, to see which strategy did best. The game had the agents interact repeatedly, simulating cooperation under conditions when it was advantageous in any given round for one agent to cheat the other.

The strategy that did the best when interacting with a variety of other strategies was very simple: On the first exchange, keep your bargain with the other agent. On subsequent rounds, do whatever the other agent did last time. This would allow the agent to cooperate, but left it free to retaliate if the other agent failed to cooperate. Such a strategy requires the agents to interact repeatedly, acquiring reputations with each other. It also requires an agent to have enough of a memory to be able to remember what the other agent did on their last encounter, which meets a minimal definition of *machine learning* in AI.

One of the virtues attributed to religion is that it encourages people to cooperate with each other (Hirschi and Stark 1969; Stark and Bainbridge 1996). It may serve this function quite apart from whether the beliefs of the religion are literally 'true', for example whether God really punishes sinners. Indeed, the human brain evolved as a useful tool for survival and reproduction, not in order that we could be scientists in pursuit of objective truth. However, biological evolution did not create religion, but only the cognitive preconditions for the cultural evolution that did so, because fully elaborated theologies, ecclesiastical institutions, and trained professional clergy did not exist until the development of urban societies based on the development of agriculture.

Axelrod's simulation is relevant to religion precisely because it achieves cooperation in the absence of supernatural beliefs. In 2008, James Dow published the results of an artificial intelligence simulation that seemed to show religion could facilitate cooperation by serving as a marker for individuals who were open to communications within a small society. However, the simulations of both Axelrod and Dow assume that pairs of agents will interact repeatedly with each other, thus gaining direct knowledge about how the other agent behaves.

Axelrod's work suggests that religion may have been unnecessary for cooperation in early human societies, organized in tiny hunter-gatherer bands, in which each person knew each other very well. Dow's work implies that religion may have begun to have a role in establishing people's reputations at the point in human history when societies began to grow beyond the small hunter-gatherer band and when many of a person's interactions would be with strangers. This suggests that our focus should move from pairs of people to whole societies, the domain of sociology.

THEORETICAL BACKGROUND

The sociology of religion is a very well-established field that dates far back into the nineteenth century, even before the term *sociology* was used to describe it. Undergraduates are often told that the founders of sociology were Emile Durkheim (1858–1917) and Max Weber (1864–1920), both of whom wrote extensively about religion. However, a better case can be made for the historical significance of Auguste Comte (1798–1857), who coined the term *sociology* and sought to replace religion with sociology, as the source of moral guidance (Comte 1883).

Comte's overarching theory is a cognitive one, that humanity has evolved culturally over time from more primitive to more advanced conceptualizations of reality. First, humans universally thought in *theological* terms that imbued nature with personality, then moved through a *metaphysical* stage that thought in terms of abstract categories rather than personalities, for example in codifying laws in a rational manner rather than relying upon Moses to bring them down from a mountaintop. Finally, with the rise of modern science, humanity adopted a *positive* approach that sought absolute knowledge though empirical research and rigorous logic. Comte divided the theological stage into three sub-stages—animism, polytheism, and monotheism—that defined the cognitive history of religion, here summarized by John Stuart Mill:

> The Theological, which is the original and spontaneous form of thought, regards the facts of the universe as governed not by invariable laws of sequence, but by single and direct volitions of beings, real or imaginary, possessed of life and intelligence. In the infantile state of reason and experience, individual objects are looked upon as animated. The next step is the conception of invisible beings, each of whom superintends and governs an entire class of objects or events. The last merges this multitude of divinities in a single God, who made the whole universe in the beginning, and guides and carries on its phaenomena by his continued action, or, as others think, only modifies them from time to time by special interferences. (1866, p. 10)

Near the middle of the twentieth century, sociologists like Talcott Parsons (1964) established a functionalist interpretation of religion and of the early sociological writers about it. Durkheim, for example, argued that religion could help provide stability to a community, reflected in a reduced suicide rate, while hinting that God was really a metaphor for society (Durkheim 1897, 1915). Weber (1930) argued that a possibly temporary *elective affinity* between Protestantism and Capitalism helped establish institutions favouring financial investment. Sociologists influenced by Parsons tended to remember Weber as saying that religion was a permanent precondition for civilization, and the Protestant variety was of special enduring value for a modern economy. The field abandoned Comte's view that theology was an outdated mode of cognition at the same time that fewer and fewer really innovative sociologists studied religion. The sociological study of religion became a subdivision of religious studies, tolerant of a range of beliefs but centred on Christianity, and uninterested in theories or research that disagreed with traditional faiths.

When cognitive scientists began writing about religion, late in the twentieth century, they tended to ignore the vast sociological literature on the topic in favour of proposing theories based on what they believed to be mental habits that individuals inherited from biological evolution. The most widely used idea in the new cognitive science literature about religion concerns the human propensity to impute conscious agency to complex events (Atran 2002). That is, we tend to assume somebody is responsible for the things that happen. This could be restated in terms of Comte's animist substage of human cognitive development. Both New Atheists and cognitive scientists tend to ignore the two subsequent substages of Comte's theological stage, which probably require sociology to explain (Pinker 1997).

Other cognitive theories also abound. Paul Bloom (2004) focuses on the delusion that humans have souls, resulting from the fact that the brain is not aware of its own operation. Porter Abbott (2003) suggests that human thought organizes things in terms of narratives—stories in which protagonists face obstacles and take actions in pursuit of goals—and thus the scientific theory of evolution cannot compete with religious stories because it is *unnarratable*. Pascal Boyer (2001) and Justin Barrett (2004) focus on a characteristic that religious beliefs have that facilitates their communication and memorization, calling them *minimally counterintuitive*. They are similar to real phenomena, so they are easily imagined and discussed, but they are different enough to be memorable and remarkable. It is also possible that belief in an afterlife reflects the fact that we develop mental models of the feelings, beliefs, and intentions of the individual people closest to us, and the models persist after the deaths of the people (Bering 2006).

Presumably, a complete theory of religion might make use of all these ideas, but they tend to focus on the cognition of individuals even when they acknowledge that the individual exists in a social environment. However, it may turn out that some of these ideas are unnecessary for the explanation of religion, including the supposed human propensity to impute conscious agency to complex events. Perhaps from early childhood, humans learn to seek help from other people when they cannot achieve their goals through individual action. The more difficult it is to attain a highly desired goal, the further outward through the social network a person will seek help. At the same time, deceptive individuals will pretend to offer help for a price—including magicians, witchdoctors, and faith healers—sometime claiming to function as intermediates with very distant help-givers who cannot be contacted directly. Religion, therefore, may have evolved not from a hypersensitive cognitive function of the individual, but from self-serving manipulation of the individual by others, initially offering crude magic but over the course of cultural evolution creating elaborate religious organizations. A rather well-developed social-cognitive theory of religion along these lines exists, sometimes called the New Paradigm in the sociology of religion (Warner 1993; Jelen 2002). It begins with seven simple axioms (Stark and Bainbridge 1987, p. 325):

Axiom 1: Human perception and action take place through time, from the past into the future.
Axiom 2: Humans seek what they perceive to be rewards and avoid what they believe to be costs.
Axiom 3: Rewards vary in kind, value, and generality.
Axiom 4: Human action is directed by a complex but finite information-processing system that functions to identity problems and identify solutions to them.
Axiom 5: Some desired rewards are limited in supply, including some that simply do not exist.
Axiom 6: Most rewards sought by humans are destroyed when they are used.
Axiom 7: Individual and social attributes which determine power are unequally distributed among persons and groups in any society.

The theory links directly to artificial intelligence, because it conceptualizes the human mind as an *information-processing system*. As in Axelrod's research, the axioms form the basis for analysis of interaction between individuals who are seeking benefits, called *rewards*, which in the New Paradigm theory may be very diverse in kind. They begin with food and security. But as human society evolved, economic markets and complex technology produced a vast catalog of goals. Especially important for effective cognition are ideas about solutions to problems, which are instrumental rewards that help the individual attain more fundamental goals. When sociologist George Homans (1974) analyzed social exchange from a perspective similar to Axelrod's, he used the example of the exchange of advice for approval, suggesting that many valued rewards are cognitive in nature.

The final three axioms on the list above introduce scarcity and competition into the picture, and lay the basis for the emergence of religion. In particular, two general goals are very difficult to achieve: (1) social status, which some people attain to the disadvantage of everybody else, and (2) eternal life, which nobody can really possess. When people cannot achieve a goal they intensely desire, they seek ideas about solutions from other people, and in the cases of high status and immortality, most people are on the same quests, so they share ideas. This can lead to the spread of what Peter Berger (1990) called *rumours of angels*, the hope that supernatural exchange partners exist who can provide status and life, perhaps with the help of human beings playing the role of intermediary. A few people may find it advantageous to claim they are in contact with the angels, and the role of priest evolves in society, a dubious profession prone to fraud and exploitation, even as it seeks to embody altruistic morality (Bainbridge 2002).

This process takes a long time, and is shaped by the cultural evolution of society, during which religious traditions compete in a manner similar to species in biological evolution. Given the diversity of individualistic cognitive science theories, the fragmentation of artificial intelligence, and the intellectual isolation of the sociology of religion, it will be difficult to bring these fields together. However, some very specific research topics in the sociology of religion can serve as entry points for artificial intelligence, and religious conversion is one of them.

SOCIAL INFLUENCE

Religious beliefs and rituals are learned during childhood and sustained through adulthood by local congregations, rather than being the spontaneous creations of each individual person. Thus the fundamental form of religious cognition is social, embodied in social networks and groups. In 1993, a workshop organized by the National Science Foundation predicted the emergence of a new field it called *artificial social intelligence* or ASI (Bainbridge et al. 1994). As with more traditional AI, progress since then has been real but slow, and one of the more active research areas uses very simple models of individual decision-making, emphasizing instead the influence of social network structures and local consensus.

A good example is a computer simulation study by Iannaccone and Makowsky (2007) of how regions of a large nation like the United States are able to preserve their religious differences, despite high rates of geographic migration between them. Specifically, the US southeastern region favours traditional forms of Protestantism that are described as fundamentalist or evangelical, whereas the western region is far more secular than the country as a whole. Inspired by early work by Thomas Schelling (1971), Iannaccone and Makowsky considered two alternative decision models: (1) individuals select a destination for their geographic migration that harmonizes with their religious beliefs, or (2) individuals move for non-religious reasons, such as employment opportunities, then adjust their personal religious beliefs to harmonize with their social environment. Both of these theories plausibly explain the persistence of regional differences, but the second provides a basis for studying how small religious minorities can persist—by creating small, cohesive communities within the wider social network—and potentially how some religious movements may grow over time, eventually changing the regional pattern.

To look more deeply into how ASI computer simulations operate, we shall use simulation software programmed by the author of this chapter, to examine how two traditional theories of social influence combine to explain much of the religious landscape (Bainbridge 2006). Called *Cyburg*, the software models a town of 44,100 people, a number chosen because it is large enough to exhibit mass effects and is convenient for simulations in which several groups with equal populations compete. The program is structured as a multi-agent system based on agent-based modelling (Epstein and Axtell 1996; Kaminka 2004; Macy and Willer 2002; Moretti 2002). An agent is an artificial intelligence computer software module that can take action in a larger computational environment, much as a human being might. Each agent represents a person, and the total system represents a society. The software is complex and easily modified, but here we shall first focus on one simple simulation for sake of clarity.

Imagine that people live in a square residential area, 210 houses on a side, and they potentially have friendship relations with their eight immediate neighbours. Depending upon how the particular simulation is set, people may or may not change their friendship ties, but with each neighbour there either is or is not a social network tie at any given point in time. At the beginning of a simulation run, the user decides how many religious groups compete within the town, without defining their beliefs and practices. One group could represent atheists, and others could be mainstream denominations. Another group could be set to proselytize aggressively, building social ties with non-members. Depending upon how the parameters are set for the simulation run, the program can model the common sociological observation that religious conversion tends to take place through personal influence from friends who already possess the particular faith (Stark and Bainbridge 1980).

Among the classical sociological ideas the program simulates is the *differential association theory* of Edwin Sutherland (1947). The term *association* is often defined by sociologists simply in terms of social connections, notably network ties to members of a particular congregation, that influence non-member friends to join the congregation. However, Sutherland was part of the symbolic interactionist movement in sociology, and often viewed associations as *mental connections between concepts*. An individual person will have learned many *definitions* of concepts, for example being aware that some people believe in God, while others do not. The question then becomes what concepts the person associates with *religious faith*—for example *stupidity* versus *respectability*. Sutherland's theory was originally framed in terms of criminal behaviour, but here are its principles stated in terms of religion:

1. Religious behaviour is learned...
2. In interaction with other persons in a process of communication...
3. Principally within intimate personal groups.
4. The learning includes:
   a. religious practices, and
   b. religious beliefs, attitudes, and values.
5. The religious beliefs, attitudes, and values are learned from definitions favourable to acceptance of the particular religion or unfavourable.
6. A person converts to a religion because of an excess of definitions favourable to the religion over definitions unfavourable to the religion.
7. Differential associations may vary in frequency, duration, priority, and intensity.
8. The process of religious learning involves all the mechanisms that are involved in any other learning.
9. While religious behaviour is an expression of general needs and values, it is not explained by those general needs and values, since non-religious behaviour is an expression of the same needs and values. (Bainbridge 2006, pp. 47–8)

Note how cognitive this theory is, while also being sociological. A more complex simulation than the one described below could incorporate the variables described in step 7, for example modelling priority by giving early messages about religion greater strength than later messages, which is how Sutherland defined the variable. Here, we will simply model how a preponderance of a particular faith among an individual's friends will influence that individual to adopt their faith. The cognition will be distributed across a large social network, so it is *social cognition*. Each individual, however, makes decisions based on rules, which also renders this a form of *rule-based* artificial intelligence, and rule-based social cognition may fall within the *automated reasoning* subfield of AI.

To achieve more realistic complexity, we will add a simple variant of Fritz Heider's (1958) *balance theory*, which is also both social and cognitive. In part, Heider restated Sutherland, arguing that a sense of cognitive equilibrium required that a person's beliefs and social attachments needed to harmonize. Having a friend who had different beliefs would cause *cognitive dissonance*

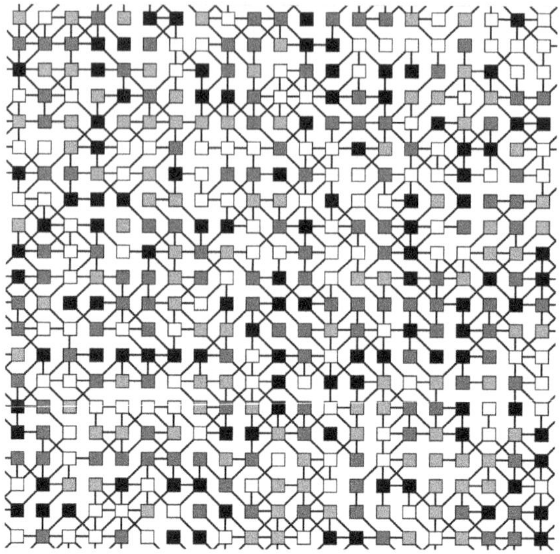

Fig. 12.1. The Cyburg Network Simulation at the beginning of one computer run

(Festinger 1957), which might impel one individual to convert to the other's faith. But balance could also be achieved by breaking off the friendship. Similarly, having a neighbour who shares one's own beliefs, but is not a friend, causes cognitive dissonance. One might resolve that different form of imbalance by building a friendship with the like-minded neighbour.

Figure 12.1 shows a small portion of the community—exactly 1% of the total population—at the beginning of a simulation run, in which membership in four religious groups of roughly equal population is distributed at random. Also at random, social bonds link about half of the pairs of neighbouring agents. Each agent is represented by a square, and we can imagine that the white squares are Atheists, the two tones of grey represent two mainstream religious denominations, like Lutherans and Methodists, while the black squares represent members of a very intense religious sect. At the beginning of this particular run, 49.96% of the possible social bonds between neighbours actually exist, and 25.32% of the population belongs to the religious sect.

The Cyburg simulation offered many options, both in setting the initial conditions and the rules that will govern the agents' behaviour. The run described here went through a series of steps. At the beginning of a turn for each agent, a random number determined whether it would follow Sutherland's theory or Heider's, with equal probabilities. Following Sutherland's theory, the agent would check each of the other agents with whom it had a social bond to see if a plurality belonged to a different religious group than the agent already belonged to. If so, the agent would convert to that group. So Sutherland's theory governed denominational membership.

The version of Heider's theory in effect governed social bonds. When it was randomly selected, the agent would check each of the neighbouring agents, breaking off a bond if the other agent belonged to a different religious group, and establishing a bond with any neighbour who belonged to the same group as the agent who was taking the turn. The one exception to this rule concerned the religious sect. On any given turn, a member of the sect has a significant chance of ignoring Heider and forming social bonds with all of its neighbours. This incorporates into the simulation a standard finding in the sociology of religion, that members of evangelical religious groups intentionally build friendships with non-members, as a mechanism for converting them to their faith (Bainbridge 1997). Of course, if a sect-member agent builds bonds with too many members of a specific other group, it may be converted away from the sect by Sutherland's theory. But over time the sect members form dense clusters in the social network, thus sustaining each other's membership while those on its social periphery build bonds with outsiders and thus convert new members.

After 20 steps through the simulation, as shown in Figure 12.2 the sect membership has reached 52.90% of the population, and the density of social bonds has increased to 62.20%. With the sole exception of their propensity to form social bonds with non-members, the sectarians differ not at all in their individual cognitions, so this simulation seems to ignore theories like the New Paradigm that assume sectarians have a special need for social status. However, a major finding of one of the early studies in cognitive dissonance is that members of a sect who are concerned about social status will tend to proselytize (Festinger, Riecken, and Schachter 1956).

As in the case of Axelrod's simulations, an interesting process of social evolution takes place despite the very constrained assumptions. After 20 steps each religious group has local clusters of members, who will support each other's faith and resist conversion. After 100 steps, the sect grew to include 76.55%, and the social bond density became 90.83%, illustrating the connection Durkheim made between religion and social solidarity. During this process, the Atheists declined from 25.15% to 15.97% to 8.10% of the population, and they wound up clustering spatially with the two ordinary religious denominations in relatively secular islands and bands across the landscape.

Many different versions of the simulation can be run, with somewhat different assumptions. The simulation just described runs forward to a point of stability, and this may accurately model the evolution of religious denominational membership, as a very small number of sects come to dominate, becoming the de facto dominant churches. The end result is what statisticians call an *absorbing state*, in which random events take the simulation into a particular configuration, from which it cannot ever escape. This suggests that conceivably

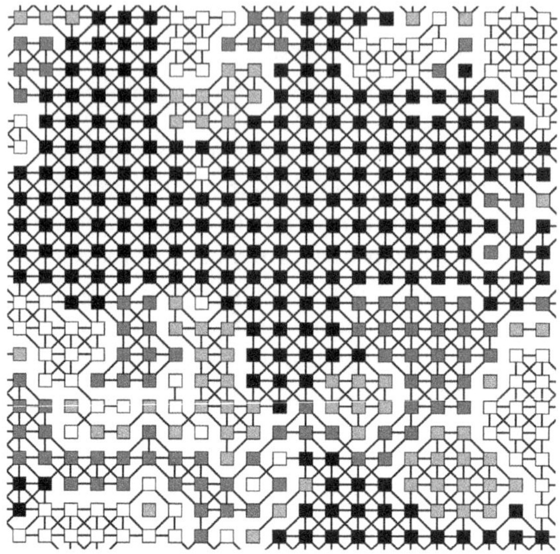

Fig. 12.2. The Cyburg Network Simulation after Step 20

the dominance of Christianity and Islam in different geographic regions of the modern world is the result of a similar, stochastic process, rather than testifying to any functional superiority of these two religious traditions. To consider such questions more deeply, we must increase the complexity of the individual cognition of the agents.

SUPERNATURAL COGNITION

Simulation of social influence across social networks typically fails to model distinctive features of the particular beliefs held by the agents applying equally well to religion, politics, or fashion fads, for example. However, a few studies do model the cognitions themselves, and the Cyburg system is capable of doing this. Among its options is a highly customizable simulation in which each of the 44,100 agents is modelled through an AI decision module that devotes over 100 memory registers to each agent, in a manner that represents machine learning of categorization schemes. It essentially allows each agent to develop a theory of the world it inhabits, and within that theory decide upon which actions to take.

Most research in artificial intelligence intends to optimize solutions to well-defined but complex problems, yet from the perspective of the New Paradigm in the sociology of religion, certainly from the New Atheist perspective, and perhaps also the perspective of cognitive science, religion is a false solution. Thus, the challenge is how to model human cognitive error, something that is easy to do but does not currently have well-established methods because the main work of artificial intelligence researchers aimed in the opposite direction, to reduce error rather than create it.

One early study modelled ethnic prejudice, using a neural net that was intentionally prone to errors, but which could often solve the problems presented to it (Bainbridge 1995a). Called Minimum Intelligent Neural Device (MIND), it modelled a classic *cognitive effort* theory of prejudice proposed by Gordon Allport (1954). It is easier for people to think in terms of gross categories, rather than fine distinctions, so they tend to lump together all members of each ethnic group other than their own. In the MIND simulation, 1000 intelligent agents interacted with members of another ethnic group, which contained a hostile subgroup that could be conceptualized as terrorists. Some of the agents were able to learn to avoid only the subgroup, but others made the prejudicial cognitive error of avoiding the entire group to which the hostile subgroup belonged, even though the majority of members in the ethnic group behaved in a benign manner.

The challenge in modelling religious faith is similar: how to design an AI that can develop various levels of irrational faith, depending upon social influence and other environmental contingencies. The Cyburg system incorporated modules that accomplished this, by first modelling how humans develop theories about the real people they interact with, and then on the same computational basis modelling the theories they may develop about supernatural beings who are similar in most respects to humans but possess the *minimally counterintuitive* quality that has become important in cognitive science theories of religion.

Originally published in simplified form a quarter century ago (Bainbridge 1987; 1995b), this AI system constantly compares several different category systems for understanding the division of labour among the agents imagined to be setting up a new society, in which a new culture must evolve. In this scenario, some agents produce energy, others produce water, others use water to produce food, and a fourth group uses both energy and water to produce oxygen. These are the four material resources required by the colonists to survive, and all are constantly consumed by the colonists. Each agent has an ID number, 1 through 44,100, and group membership is assigned in order. For example, suppose the particular computer run defines four groups in the population, and the third group produces food. Then the IDs of food producers will be a sequence beginning 3, 7, 11, 15, 19, starting with 3 and adding 4 each step. Of course the human mind employs much more complex categorization methods, but as a metaphor this simple mathematical algorithm does model an essential component of the theory.

The agents constantly interact with each other, seeking exchanges, and after a successful exchange both agents add a number to the memory registers that represent the exchange, each from its own viewpoint, and applying multiple theories in parallel. One agent interacts with a member of group three, and receives food. The number in that agent's food memory register for 'group three of four groups' will be increased, which also strengthens the agents 'faith' in that theory. When selecting an exchange partner, an agent will go through a two-step process. First, it will decide which classification theory to follow, in the examples here whether the society consists of two, three, or four groups. Second, it will decide which group within the selected classification scheme is most likely to provide the desired resource. The classification theory is decided by comparing the success the competing theories have had in the past in identifying exchange partners, measured as the statistical variance across the memory registers for that theory.

Setting aside the technical details, this system can indeed quickly learn how to obtain the necessary rewards, and the entire simulated colony settles into an economically viable state. However, the agents grow older, begin to worry about where they can get a fifth resource—extended life—and begin seeking other agents who can provide it. Here is where the human tragedy and religion enter the picture. No agents represented within this simulated world can in fact give extended life to each other. As time passes, the agents become more frantic, seeking life more and more of the time. As they are repeatedly disappointed in their interactions with each other, the numbers in the corresponding memory registers are reduced.

At the beginning of a run, each memory register is filled with a random integer in the range 500-999. Every such disappointment reduces that number a little, to an arbitrary minimum of 50 which represents the least hope that an agent can have in any theory. However, some memory registers represent agents that do not exist in the simulation, such as 'group four of three groups'. This is a supernatural category that does not exist in any realistic numbering system. Because an exchange cannot actually be completed with such an agent, the number in the memory register representing it does not get reduced, and thus gains strength in competition with the more naturalistic theories whose memory registers now contain much smaller numbers.

Granting that this is a mathematical metaphor, rather than an accurate statement of how human cognition actually works, it may be a good one. High school algebra students are taught to distinguish between *real numbers* and *imaginary numbers*, but accord some respect to both. 'Group four of three groups' is a minimally counterintuitive *supernatural number*. It is quite possible to do calculations with it, but it violates one rule of the *natural number* system: The remainder from dividing integer A by integer B must be less than B, and only a supernatural number could escape this rule.

Table 12.1 shows the result of five runs of the simulation, with slightly different starting conditions. In these runs, each agent compares three different theories about the societal division of labour—that people are divided into

Table 12.1. Percentage of hope invested in supernatural agents, in five AI simulation runs

Simulation Step	Without any communication	With communication about life			
		0 Messiahs	1 Messiah	3 Messiahs A	3 Messiahs B
0	25.00	25.00	25.00	25.00	25.00
20	30.88	49.44	56.22	64.79	71.76
40	34.93	71.01	83.00	75.52	91.35
60	36.82	59.72	93.23	89.15	93.00
80	38.10	60.02	93.37	93.97	93.04
100	39.45	60.37	93.42	94.05	93.03

(1) two groups, or (2) three groups, or (4) four groups—without assuming that the same theory applies to all desired resources. They do not initially know that the simulation specifies there are four groups, each specializing in one of the four physical resources, and that no people in the society can provide life. All five runs begin with a 25% chance that an agent will seek extended life from a supernatural agent, which means an agent belonging to one of these three (non-existent) groups: (1) the third of two groups, (2) the fourth of two groups, or (3) the fourth of three groups.

The first run simply has the agents interact, doing nothing more than exchanging with each other. Their tendency to seek life from supernatural beings increases over time, but only slowly, as they individually learn they cannot obtain it from each other. In the second run, when two agents have a successful exchange of material resources, they also discuss life—which means that the one with slightly more conviction in a theory will influence the one with lower conviction to adopt their views. Indeed, in the second column of the table we see that faith increases much more rapidly than without such communication. What happens is that the opinions of the most extreme agents spread to the others. The third run in the table began by selecting one single agent, and adding 2000 to the memory register for the fourth of three groups, making this person a Messiah promulgating faith in a particular deity. Notice that this person's doctrine comes to dominate the culture quickly in the absence of any opposition. The two final runs start identically with three Messiahs, each with a different faith, and show that considerable variation can occur in the speed of early growth, even as the final result is about the same.

The first simulation run represents purely individual development of superstitions, of an animist sort, in which most individuals come to rely more and more upon supernatural beliefs, but usually without a strong faith in any one god. The other runs illustrate two of the three primary theories of cult formation, essentially how all religions begin (Bainbridge and Stark 1979). In the *subculture evolution* theory, illustrated by the second column, individuals share their hopes about supernatural help, gradually adopting the indistinct views of the most extreme members of the society, but again without a well-focused belief in one god. The three concluding columns illustrate the *psychopathology* theory of cult formation, in which one or more rare individuals experience the equivalent of psychotic episodes, in which they learn radical ideas (here injected by the experimenter) unrelated to their real-world experience (Stein and Lidik 1998), then share them with other people while playing the role of Messiah.

Cyburg does not model the third *entrepreneur* theory of cult formation, but Afzal Upal (2005) has taken the first steps in doing so. A religious entrepreneur assumes the role of Messiah because it is profitable to do so, receiving tangible rewards from followers. To a first approximation, there are two kinds of religious entrepreneurs, honest ones and dishonest ones. Upal modelled this distinction by having some Messiahs promulgate behaviour rules they had personally found effective, and others promote ineffective behaviour rules. In future, it ought to be possible to create much more complex artificial intelligence models of the mind, allowing an agent to develop two theories: (1) *effective* ones they found worked for them in their ordinary lives, and (2) *ineffective* ones they found it especially easy to sell to other people. Effective rules for attaining immortality do not exist, but it is still possible for an agent to believe a particular rule is effectively improving the life span on the basis of random events or cognitive error. Whether there really is a fundamental difference between honest and dishonest Messiahs, remains to be seen.

CONCLUSION

The artificial intelligence simulations described here model the emergence of religion as a process of evolution in social cognition, and in some cases the persistence of multiple faiths in a form of dynamic equilibrium. Thus they roughly illustrate Comte's theological stage of human cultural evolution, but not the later stages he theorized in which traditional faiths would vanish from the scene. Close examination of the spiritual cognition simulation performed in the Cyburg system, however, indicates that animism and polytheism would survive the emergence of monotheism, because the AI agents always had some tendency to explore supernatural alternatives, and some end-states of a simulation were decidedly polytheistic, depending upon the initial conditions and accidents of history. Traditional religion remains socially significant, many years after Comte wrote, which would seem to disprove his theory.

We noted that religion may be an absorbing state, into which a simulation or a real society can wander, but from which it cannot escape. This actually models a major sociological theory of religion different from Comte's and proposed by early twentieth-century writers such as Oswald Spengler (1926) and Pitirim Sorokin (1937–1941). In their somewhat different formulations, each major civilization develops an essentially religious orthodoxy early in its history, that prevails until the civilization falls. During a Dark Age of extreme social disorganization, faith may be erased, setting up conditions for a rerun of the evolution of a religion. Any of the AI simulations can model that, simply by rebooting to a new random set of initial conditions after a while.

It is possible, however, for evolution to open an escape route from an absorbing state, without the necessity for total collapse of the system. Note that the supernatural cognition simulation described above did not assume that the same categorization theory applies to all desired resources. Suppose it did, and over time each agent would gradually learn a categorization theory that applied to all five rewards: energy, water, food, oxygen, and life. Then the agents would wind up believing that there were exactly four groups of other agents they could interact with, because for most of the five rewards this was the most rewarding theory. Since there really were four groups of agents in the simulation, this would leave them in the end with no faith in the supernatural, and little hope that any exchange partner could provide them with eternal life—which is the sad truth of their world. This modification to the simulation would model Comte's theory, that over time science would come to answer all questions that have answers, in terms of one comprehensive cognitive system, and bring the Theological Era to a decisive conclusion.

The views expressed in this essay do not necessarily represent the views of the National Science Foundation or the United States. [Legally required disclaimer]

REFERENCES

Abbott, H. P. 2003. Unnarratable knowledge: The difficulty of understanding evolution by natural selection. In *Narrative Theory and the Cognitive Sciences*. Ed. D. Herman. Stanford, CA: Center for the Study of Language and Information, 143–62.

Allport, G. 1954. *The Nature of Prejudice*. Boston: Beacon.

Atran, S. 2002. *In Gods We Trust: The Evolutionary Landscape of Religion*. Oxford: Oxford University Press.

Axelrod, R. M. 1984. *The Evolution of Cooperation*. New York: Basic Books.

Axelrod, R. and W. D. Hamilton. 1981. The evolution of cooperation. *Science 211*: 1390–6.

Bainbridge, W. S. 1985. Cultural genetics. In *Religious Movements*. Ed. R. Stark. New York: Paragon, 157–98.

——. 1987. *Sociology Laboratory*. Belmont, CA: Wadsworth.

——. 1995a. Minimum intelligent neural device: A tool for social simulation. *Mathematical Sociology 20*: 179–92.

——. 1995b. Neural network models of religious belief. *Sociological Perspectives 38*: 483–95.

——. 1997. *The Sociology of Religious Movements*. New York: Routledge.

——. 2002. A prophet's reward: Dynamics of religious exchange. In *Sacred Markets, Sacred Canopies*. Ed. T. G. Jelen. Lanham, MD: Rowman and Littlefield, 63–89.

——. 2006. *God from the Machine: Artificial Intelligence Models of Religious Cognition*. Walnut Grove, CA: AltaMira.

——. 2009. Atheism. In *The Oxford Handbook of the Sociology of Religion*. Ed. P. B. Clarke. Oxford: Oxford University Press, 319–35.

——. 2010. Cognitive science and the new atheism. In *Religion and the New Atheism*. Ed. A. Amarasingam. Leiden: Brill, 79–96.

——. 2012. Artificial intelligence. In *Leadership in Science and Technology*. Ed. W. S. Bainbridge, 464–71. Los Angeles: Sage.

Bainbridge, W. S., E. E. Brent, K. M. Carley, D. R. Heise, M. W. Macy, B. Markovsky, and J. Skvoretz. 1994. Artificial social intelligence. *Annual Review of Sociology 20*: 407–36.

Bainbridge, W. S. and R. Stark 1979. Cult formation: Three compatible models, *Sociological Analysis 40*: 285–95.

Barrett, J. L. 2004. *Why Would Anyone Believe in God?* Walnut Creek, CA: AltaMira.

Berger, P. L. 1990. *A Rumor of Angels*. New York: Anchor Books.

Bering, J. M. 2006. The cognitive psychology of belief in the supernatural. *American Scientist 94*: 142–9.

Bloom, P. 2004. *Descartes' Baby: How the Science of Child Development Explains what Makes Us Human*. New York: Basic Books.

Boyer, P. 2001. *Religion Explained: The Evolutionary Origins of Religious Thought*. New York: Basic Books.

Comte, A. 1883. *The Catechism of Positive Religion*. London: Trübner.

Dawkins, R. 2006. *The God Delusion*. Boston: Houghton Mifflin.

Dennett, D. C. 1995. *Darwin's Dangerous Idea*. New York: Simon and Schuster.

Dow, J. 2008. Is Religion an Evolutionary Adaptation? *Journal of Artificial Societies and Social Simulation 11*(2). Available online at <http://jasss.soc.surrey.ac.uk/11/2/2.html>.

Durkheim, E. 1897. *Suicide*. Glencoe, Illinois: Free Press.

——. 1915. *The Elementary Forms of the Religious Life*. London: G. Allen and Unwin.

Epstein, J. M. and R. Axtell. 1996. *Growing Artificial Societies: Social Science from the Bottom Up*. Washington, DC: Brookings Institution Press.

Festinger, L. 1957. *A Theory of Cognitive Dissonance*. Evanston, Illinois: Row, Peterson.

Festinger, L., H. W. Riecken, and S. Schachter. 1956. *When Prophecy Fails*. Minneapolis: University of Minnesota Press.

Heider, F. 1958. *The Psychology of Interpersonal Relations*. New York: Wiley.

Hirschi, T. and R. Stark. 1969. Hellfire and delinquency. *Social Problems 17*: 202–13.

Homans, G. C. 1974. *Social Behavior: Its Elementary Forms*. New York: Harcourt, Brace Jovanovich.

Iannaccone, L. R. and M. D. Makowsky. 2007. Accidental Atheists? Agent-Based Explanations for the Persistence of Religious Regionalism. *Journal for the Scientific Study of Religion 46*: 1–16.

Jelen, T. G. Ed. 2002. *Sacred Markets, Sacred Canopies*. Lanham, MD: Rowman and Littlefield.

Kaminka, G. A. 2004. Multi-Agent Systems. In *Encyclopedia of Human-Computer Interaction*. Ed. W. S. Bainbridge, 475–80. Great Barrington, MA: Berkshire.

Macy, M. W. and R. Willer. 2002. From factors to actors: Computational sociology and agent-based modeling. *Annual Review of Sociology 28*: 143–66.

McCorduck, P. 2004. *Machines Who Think*. Natick, MA: A. K. Peters.

Mill, J. S. 1866. *Auguste Comte and Positivism*. London: N. Trübner.

Moretti, S. 2002. Computer simulation in sociology. *Social Science Computer Review 20*(1): 43–57.

Parsons, T. 1964. Evolutionary universals in society. *American Sociological Review 29*: 339–57.

Pinker, S. 1997. *How the Mind Works*. New York: Norton.

Russell, S. J. and P. Norvig. 2003. *Artificial Intelligence: A Modern Approach*. Upper Saddle River, NJ: Pearson.

Schelling, T. C. 1971. Dynamic models of segregation. *Journal of Mathematical Sociology 1*: 143–86.

Sorokin, P. A. 1937-1941. *Social and Cultural Dynamics*. New York: American Book Company.

Spengler, O. 1926. *The Decline of the West*. New York: Knopf.

Stark, R. and W. S. Bainbridge. 1980. Networks of faith: Interpersonal bonds and recruitment to cults and sects. *American Journal of Sociology 85*: 1376–95.

——. 1987. *A Theory of Religion*. New York: Lang and Toronto Studies in Religion.

——. 1996. *Religion, Deviance and Social Control*. New York: Routledge.

Stein, D. J. and J. Lidik. Eds. 1998. *Neural Networks and Psychopathology*. Cambridge: Cambridge University Press.

Sutherland, E. H. 1947. *Principles of Criminology*. Philadelphia: Lippincott.

Upal, M. A. 2005. Simulating the Emergence of New Religious Movements. *Journal of Artificial Societies and Social Simulation* 8(1). Available online at <http://jasss.soc.surrey.ac.uk/8/1/6.html>.

Warner, R. S. 1993. Work in progress toward a new paradigm for the sociological study of religion in the United States. *American Journal of Sociology 98*: 1044–93.

Weber, M. 1930. *The Protestant Ethic and the Spirit of Capitalism*. New York: Scribner.

Whitehouse, H. 1995. *Inside the Cult: Religious Innovation and Transmission in Papua New Guinea*. New York: Oxford University Press.

13

Concluding Reflections

MALCOLM JEEVES

Over the past decade one of the leading researchers in the field of the cognitive science of religion (CSR) has been Justin Barrett. He sees the cognitive science of religion as 'the scientific study of religious thoughts and actions, which draws upon the cognitive sciences and evolutionary psychology'. CSR scholarship, he says, 'has begun to sound off on what cognitive mechanisms humans must have to conceptualize a god and to generate actions for interacting with that god'. Recurrently, HO-ToM (higher order—theory of mind) appears as a central player in these accounts. Barrett has recently identified, 'a few features of the CSR account of religion that merit highlighting'. It is worth reminding ourselves of these as we consider whether they adequately do justice to the full diversity and complexity of religion. Barrett (in press) writes:

> First, these cognitive accounts typically presume strong biological and cognitive continuity with ancestral species and, by extension, with non-human primates through to approximate human ancestors. Second, though conceding the possibility of later exaptation, cognitive accounts at their core are evolutionary byproduct accounts of religion. That is, the cognitive equipment that gives rise to religious expression is presumed to have evolved under selection pressures unrelated to religion or religious entities. Whereas the ease with which humans acquire fear of snakes presumably evolved in response to snakes themselves as a survival threat, the ease with which humans acquire belief in gods is not thought to have evolved in response to gods. Under these accounts religious thoughts are an evolutionary byproduct and not an adaptation *per se*. Third, many different cognitive sub-systems or 'mental tools' cooperate to encourage religious beliefs and practices, and so belief in gods (or souls, the afterlife, etc.) are byproducts of multiple cognitive adaptations and not just one. Nevertheless, one 'mental tool' takes center stage in these accounts: ToM.

There is no doubt that the approach of Justin Barrett, and those like him, has been extremely fruitful. However, any approach which has the label 'cognitive science of religion' should be carefully scrutinized to examine the extent to which it draws upon the wide range of disciplines covered by the label 'cognitive science'. Perhaps this book should have started with a quote from the last chapter by William Bainbridge, who has reminded us that one of the leading publications in the field of cognitive science—*Cognitive Science: A Multidisciplinary Journal*—lists seven disciplines that came together to create the new field of cognitive science. These were artificial intelligence, education, linguistics, neuroscience, philosophy, psychology, and anthropology. It is a fair question to ask whether overconcentration on the undoubted successes of the research strategies used by researchers like Justin Barrett may have resulted in insufficient attention being paid to insights into the origins of religion provided by other disciplines that contributed to the new field of cognitive science. Perhaps surprisingly, in that list there is no mention of evolutionary biology, and no mention specifically of evolutionary psychology, both of which figure prominently in 'typical' CSR accounts such as Barrett's.

The history of science illustrates over and over again how advances are made when widely accepted theories are challenged and their adequacy called into question. Over the past two decades many new insights into understanding the evolution of religion have undoubtedly emerged as a result of various cognitive science of religion approaches to the study of religion. Now, as in other areas of science, the time has come to ask whether we may have become too satisfied with the adequacy of the most widely accepted CSR approaches, and whether they can, in fact, adequately answer many central questions about the emergence of religion, or do justice to religion in all its historical diversity and complexity. This book has made an important start to that reassessment.

It is timely, then, to step back, take a critical look at the important contributions of conventional CSR approaches, and to consider their strengths and weaknesses, as well as possible alternatives. Two questions might be asked. First, how should we evaluate the contributions thus far of CSR approaches to our understanding of the evolution of religion? Second, are there new leads from a reassessment of the present status of CSR which may guide fruitful future research into our understanding of the emergence and development of religions?

At the start of the book Pyysiäinen helpfully reminds the reader that the cognitive science of religion, 'draws not only from cognitive science but also from evolutionary and developmental psychology, neuroscience, computer science, ethology and anthropology'. He also reminds us that, 'as an attempt to interpret historical data or to theorise about religion at a general level, the cognitive science of religion is *one perspective* on the study of religion or comparative religion' (my italics). It is this reminder, that CSR is only one of many possible approaches, which becomes a pervasive theme in what follows. Recognizing that there is more than one approach, Pyysiäinen underlines the fact that, 'cross fertilization of ideas is extremely important because the traditional academic boundaries between disciplines often make truly interdisciplinary work difficult and because religion is seen as a heterogeneous category and can be approached at different levels'. He notes further, and very importantly, as regards the future, that, 'an optimal strategy in research is thus a multi-level mechanistic explanation that specifies the various sorts of mechanisms (neural, cognitive, etc.) that produce and support religious phenomena at various ontological levels'.

Michael Ruse follows this warning from Pyysiäinen, reminding us that when a particular approach becomes popular it is all too easy to assume that everyone is singing from the same hymn sheet. Rather, Michael Ruse notes, 'the one thing that strikes a disinterested observer is the extent to which there is a total lack of uniformity by enthusiasts'. The result, he says, is that, 'one does not have established theories but rather suggestions based on successes elsewhere'. 'This,' he writes, 'is very much how I see present-day work on the evolution of religious beliefs.' The reference to 'successes elsewhere' refers to such things as cognitive scientists' widely accepted way of thinking in terms of modules. Its present limitations, however, should not, according to Ruse, discourage future research. He notes, for example, that, 'there is today a range of suggestions about the possible evolution of religious states. Not all of them are particularly well taken, but if one takes seriously evolutionary biology then, at the least, we should allow the possibility that one or more of the proposals will develop into deeper explanations of religious beliefs'. He wisely adds, 'whether or not any of this has implications for the truth status of religion is another matter. As a general rule, probably it does not.'

Continuing the theme of the need to be alert to the limitations of any one approach, such as CSR, Aku Visala comments, after his extensive review of the evidence, that, 'it seems that the religious relevance of cognitive and evolutionary approaches to religion are (at least in the case of theism) relatively neutral. In other words, when we think about the probability of theism as opposed to atheism, CSR gives us very little reasons to prefer atheism over theism'.

There will doubtless be many changes to the various CSR approaches which develop, and some will afford a deeper understanding of the origins and functions of religion, giving new insights into potential benefits for religious praxis at both an individual and a group level. These may include opening up fresh ways of helping to resolve ongoing tensions and divisions between religions, whether at the level of the individual or the group. It will be helpful to remember the importance of recognizing the different expressions of religious beliefs at both individual and corporate levels. All of that must be a big plus coming from such research, but none of it will, in itself, as Visala implies, enable easy decisions to be made about the truth status of specific religious beliefs. For that there will be no short cut, and the truth status of key beliefs in different religions must still be put under the spotlight of the work of historians, philosophers, theologians, anthropologists and archaeologists.

Assuming a positive approach to the fruits thus far of CSR, Purzycki and his colleagues urge us to consider that whilst this may be *part of the answer* to the question of how world religions appeared in human history, nevertheless *it may not be the full story*. Indeed, they believe that these kinds of account alone cannot explain a number of key concepts often encountered in religious systems. In order to do this they believe that these approaches must be supplemented by what they call an adaptationist account of the emergence of religion. Thus they write, 'adaptive accounts of religion do not disagree with the central and often empirically supported claim of by-product accounts that religious concepts are made possible by evolved psychological mechanisms, which evolved as a consequence of selective pressures that are in all probability unrelated to religion'. However, they add, an adaptive approach goes a few steps further. They believe that for a fruitful understanding of the emergence of religion, what they call a 'dynamic systems approach' is required. This will shed more light on all religion's components, and specifically on 'how the content of these components changes in accordance with how people interact with their social and ecological environments'. And hence their very strong argument is that, 'only by giving equal attention to the nature of the human mind, the natural and social environments, history, and all the forces which exert influence upon them and their relationships will we be able to come to terms with explaining why humans do the things they do'.

Nathaniel Barrett brings out into the open what some other writers have hinted at, namely, there is more than one way to approach attempts to understand the emergence and origins of religion and religious beliefs. He notes specifically that, 'Most cognitive-evolutionary theories claim that the propensity for supernatural belief was in fact selected for, directly or indirectly, either because it was adaptive in ancestral environments or because it emerged as a by-product of inferential mechanisms that were adaptive'. But Barrett challenges this widespread approach and states instead that, 'I would like to focus on how the turn to value reverses the priorities of evolutionary explanation. For the standard cognitive evolutionary approach, the experience of value is not really a part of cognition but something added to it, a motivational overlay. You might say that value is the icing on the cake, which is the generation of belief. In contrast, for the axiological approach that I am proposing, the priority is reversed: experience of value is the cake, and belief is the icing.'

Fraser Watts' contribution is outstanding because it no longer speaks in terms of generalities but gives a specific example of how commitment to a particular theory in cognitive psychology may become so dominant that other possibilities, equally valid on the available evidence, are not given due consideration. He argues for recognizing *evolutionary developments in the central cognitive architecture of human beings alongside any talk of modules and devices*, and, moreover, he questions the distinction between agents and non-agents. He argues that evolutionary CSR is unnecessarily committed to speculative assumptions that constrain its theorizing. Watts very wisely reminds us that, 'which choice one makes in framing scientific theories is often influenced by background metaphysical assumptions, and they are likely to be relevant to whether people prefer the view that there was initially a clear, intuitive distinction between agents and non-agents that was somewhat obscured through hypersensitive agency detection, or the view that the intuitive distinction between agents and non-agents emerged only gradually. The latter view, if it turns out to be correct, would be more congenial from a religious point of view, because it avoids the assumption that religion is some kind of cognitive mistake.' And he later develops this view, adding perceptively, 'It seems to me that recent cognitive science of religion has, in effect, been developing a competing myth of a "fall" from naturalism. There is clearly a possibility of contemporary scientists, who hold naturalistic assumptions, projecting those assumptions back onto emerging humanity.'

The danger of tunnel vision by overly strict adherence to traditional CSR theories is further underlined by Van Huyssteen. He points out that, 'The question of the evolution of religion and religious behaviour *can never be disentangled from the evolution of embodied human personhood*'. This is a timely and perceptive comment. The understanding of the emergence of human personhood is a very actively researched area at the present time, and CSR theories must do justice to developments in this area. In developing his argument for foregrounding the central theme of embodied human personhood, he draws upon the work of neuroscientists such as Patrick McNamara on the understanding of the brain sites involved in religious experience, and also the work of Wesley Wildman, whose views he quotes with approval. He writes, 'Wildman thus wants to present a religious anthropology by focusing on the embodied *Homo religious*, and by enlarging the scope of religious behaviours, beliefs, and experiences, to encompass everything relevant to human meaning and value'. He continues, 'Finally he (Wildman) developed his own, and in my view most plausible view, that religion in evolutionary terms is a combination of side-effects of both adaptive and non-adaptive features of the human organism'.

Another critical comment made by Van Huyssteen concerns the relative status of this or that scientific explanation of religion. He sums up his own view by quoting, with approval, the views of Jeffrey Schloss and Michael Murray who write, 'The different models that are now used to explain religion show not one thing more than that we have certain mental tools (perhaps selected, perhaps spandrels) which under certain conditions give rise to the belief in the existence of entities which tend to rally religious commitments. Pointing that out, however, does nothing, all by itself, to tell us whether those religious beliefs are justified or not. The mere fact that we have beliefs that spring from mental tools selected for by natural selection is, all by itself, totally irrelevant to the justification of beliefs that spring from them.'

Van Huyssteen's concluding comments on the complementarity of different approaches to the understanding of the development of religion lead directly into some of Léon Turner's major concerns. Van Huysteen wrote, 'the theological implications of this debate are that various scholars have argued that cognitive approaches and adaptationist approaches can in fact be seen as complementary'. However we are immediately warned by Turner that any too-easy acceptance of the notion of complementarity can be challenged. He argues, for example, that the CSR approach frequently courts conflict with the humanities in their (CSR's) pursuit of naturalistic causal explanations of religious phenomena which implicitly or explicitly deny the need for, or the value of, social and historical accounts of these phenomena. Focusing primarily on the work of Justin Barrett, Turner argues that, 'some CSR theorists, including Barrett, seem comfortable with naturalistic causal explanations both of the origins of belief in gods, and the pan-cultural recurrence of certain beliefs that explicitly and deliberately preclude the possibility of their explanation in terms of meaningful sociohistorical narratives, and therefore overtly conflict with a fundamental principle of (at least some) humanistic and social-scientific accounts'. Finally, Turner concludes that, 'accounting for the origins of religious beliefs in gods and accounting for the origins of the natural tendency to interpret certain ambiguous objects and events in agentic terms (which may support or reinforce belief in gods) are quite different enterprises. A causal explanation of the former in terms of the latter is potentially problematic for the humanities and social sciences.'

Timothy Jenkins is another contributor who underlines the dangers of any approach to the understanding of the development of religion which focuses exclusively on CSR accounts. Thus he asks us, 'to consider cognitive scientific discussions of religion from an anthropological perspective'. He writes, 'my suggestion is that contemporary evolutionary theory no longer assumes a more or less uniform genetic profile to a vast population united by common descent; instead it emphasizes small populations interacting in different environments with a variety of partners so that (to offer an imaginary instance) in one valley, orchids and wasps are involved in symbiosis, in another interaction is with other species of insect—flies, beetles or ants—and in a third competition with another plant species shapes local outcomes'. The relevance of this to his main argument is that, 'just as the study of genetic expression has moved from gene to protein, and that of human capacities has moved from brain to mind, so the study of evolutionary processes has moved from blind uniform populations, defined by common descent, to small heterogeneous populations involved in diverse local alliances across species (or against nature). In a word, naturalism is becoming historicised.' In contemplating the possible reconciliation of CSR's naturalistic approach to religion with those favoured by anthropologists, he concludes, '... the reconciliation will be through making nature historical, rather than making history natural'.

After a series of searching critiques of some aspects of contemporary CSR approaches to the study of religion, we turn to the views of Newson and Richerson, who largely endorse evolutionary approaches such as CSR. Their contribution is significant both because they are anthropologists knowledgeable in detail about Darwinian theory, and because they attempt to deal with a specific set of issues that arise in modern times. Hence the title of their chapter: 'Religion: The Dynamics of Cultural Adaptations'.

These authors accept that Darwinian theory, 'should be thought of as a large tool-kit of models and empirical approaches which can be deployed flexibly to solve a wide range of problems'. They note further that this tool-kit, 'is useful whenever a problem involves transmitted information changing over time'. They therefore use this approach, 'to look at the pattern of cultural change sometimes referred to as modernization, which is associated with economic development'. They note, 'these changes include changes in religious belief'. More specifically, 'they review the evidence that the size and connectedness of a population determines the complexity of the culture that a population can support'. With this in mind they look at such changes over the last 10,000 years, particularly in societies that have undergone economic development, noting, 'the "great leap forward" in cultural complexity of the Upper Paleolithic occurred because the peoples inhabiting Europe and Asia found it possible to maintain a sufficiently large communicating population'. They also note that, 'shared religious beliefs and ceremonies may have been among the cultural mechanisms these people used to facilitate the sharing of cultural information between large numbers of people dispersed over a large area'. Thus, for them, when taking a Darwinian approach to understanding human behaviour, it is as important to take account of the social networks of our ancestors living in preindustrial communities as it is to take account of the evolved cognitive modules so salient in CSR approaches.

As noted above, the final chapter comes as a timely reminder of the wide range of disciplines included today in definitions of cognitive science. William Bainbridge, an expert in artificial intelligence (amongst other things), makes a compelling case for recognizing the potential contribution of that field of knowledge to the understanding of the origins and functions of human religion. In taking a particular example Bainbridge argues convincingly for the way in which an AI modelling approach might lead us to a fresh understanding of, for example, the evolution of religious denominational membership. This is another reminder, and an important one, that there is a wide variety of perspectives required to begin to do justice to the diversity and complexity of religion. Focusing on CSR alone is not enough.

LOOKING BACK AND LOOKING FORWARD

Recently Peter Higgs, co-discoverer of the Higgs Boson, the focus of much recent scientific discussion, was asked what he thought and hoped the establishment by the University of Edinburgh of the Peter Higgs Centre for Theoretical Physics might achieve. He answered, 'I hope it will bring a kind of breadth of theoretical physics into the future of the School of Physics and Astronomy. *From my personal experience I think it's a good thing for people working in a particular subfield of theoretical physics to be aware of what's going on in other areas*.'

The brief overview above has underlined in this field the need '*to be aware of what's going on in other areas*' and a greater awareness of how the potential contributions from the disciplines subsumed under 'cognitive science' can lead to a much enriched and self-critical assessment of where we have reached so far.

Returning to our opening two questions of how to evaluate CSR approaches and whether there are promising new leads identifiable, we may summarize our findings by noting four pointers to the future.

First, there is both strength and depth, here, in the areas of scholarship involved in various subfields of the cognitive science of religion. There is both uniformity and diversity of approach. Thus, several contributors focused on what one might call a standard accepted version of CSR. Others, such as Fraser Watts, wished to provoke a rethinking, based on a focus on central processing mechanisms rather than modules. William Bainbridge, on the other hand, demonstrated how a totally different approach (namely, the application of artificial intelligence to the study of religion) may not only prompt us to reconsider the extent to which the standard model of CSR is relied upon, but also open up fresh pathways for future research.

Secondly, there is a need to encourage further direct interactions and cross-referencing between leaders in the various specialist areas covered. This will afford certain necessary checks and balances, and alert to the ever-present danger of giving way to the temptation to overly depend on the findings from only one approach. Such overdependence can so easily sweep under the carpet key data and important questions which arise from the various different perspectives that are possible upon the same data.

Thirdly, it is interesting to note a number of recurrent themes which crop up in almost every contribution:

- there are multiple possible perspectives (Bainbridge, Jenkins, Turner, Purzycki et al., Van Huyssteen, Watts).
- that different people bring different presuppositions (Barrett, Ruse, Turner, Pyysiäinen, Watts).

- that any particular phenomenon may be fruitfully investigated at several different levels (Jenkins, Turner, Pyysiäinen).
- that there is an ever-present temptation to fall into unthinking reductionism (Pyysiäinen,Turner).
- that one must always distinguish between the validity of any particular model that is being put forward about the origins and functions of religion, and the truth status of any assertions that are made within a particular religion (Ruse, Visala).

Finally, these particular contributions to the understanding of mechanisms that are supposed to account for the emergence and development of religion will undoubtedly be revisited as the twenty-first century evolves. After one or two decades have passed, it will be possible to see how firmly some of the views expressed here have become established, how others have been significantly modified, and how others, in the light of accumulating empirical evidence, no longer seem tenable.

REFERENCES

Barrett, J. L. and M. J. Jarvinen. In press. Evolutionary byproducts and *Imago Dei*. In *The Emergence of Personhood: A Quantum Leap?* Ed. M. Jeeves. Grant Rapids: William B. Eerdmans.

Subject Index

Author Index